THE COMPLETE GUIDE TO

HOME PLUMBING REPAIR AND REPLACEMENT

A Practical Guide to (Almost Always) Doing It Yourself

R. DODGE WOODSON

BETTERWAY BOOKS
Cincinnati, Ohio

Other books by R. Dodge Woodson:

The Complete Guide to Buying Your First Home
Get the Most for Your Remodeling Dollar
Rehab Your Way to Riches

Cover photograph by Susan Riley
Design & Typography by Studio 500 Associates

The Complete Guide to Home Plumbing Repair and Replacement. Copyright © 1992 by R. Dodge Woodson. Printed and bound in the United States of America. All rights reserved. No part of this book may be reproduced in any form or by any electronic or mechanical means including information storage and retrieval systems without permission in writing from the publisher, except by a reviewer, who may quote brief passages in a review. Published by Betterway Books, an imprint of F&W Publications, Inc., 1507 Dana Avenue, Cincinnati, Ohio 45207. 1-800-289-0963. First edition.

96 95 94 93 92   5 4 3 2 1

Library of Congress Cataloging-in-Publication Data

Woodson, R. Dodge (Roger Dodge)
   The complete guide to home plumbing repair and replacement : a
  practical handbook to (almost always) doing it yourself /
R. Dodge
  Woodson.
      p.   cm
    Includes index.
    ISBN 1-55870-248-2 (pbk.) : $16.95
    1. Plumbing—Amateurs' manuals.   I. Title.
TH6124.W669 1992
696'.1—dc20
                                    92-17000
                                        CIP

This book is dedicated to my wife, Kimberley, and my daughter, Afton. Their love and support are strong motivation in all my endeavors. Words cannot describe my appreciation for the two most important people in my life, Kimberley and Afton.

A special thanks is due to Ken Luce. Ken has provided support and has helped to cover the bases of the day-to-day business world while I wrote this book. Thanks, Ken.

No dedication of mine would be complete without thanks to my parents, Maralou and Woody. My parents have helped me reach my goals for many years and they are still helping. Thanks, Mom and Dad.

Acknowledgments

I would like to thank Andrew Wallace for his contribution to this book as a photographic model.

I would like to thank the following companies for providing photographs and illustrations to make this book more effective and useful.

A.O. Smith Corp.

A.Y. MacDonald Mfg. Co.

American Standard Inc.

Amtrol Inc.

CR/PL, Inc.

Fernco, Inc.

General Wire Spring Co.

Gould Pump Inc.

Hellenbrand Water Conditioners

Kohler Company

Moen Inc.

Ridge Tool Co.

UNR Home Products

Universal-Rundle

Vanguard Plastics, Inc.

Zoeller Co.

Contents

Introduction

If you have ever wished one of your family members was a plumber, this book is the next best thing. Written by a master plumber with over seventeen years of residential experience, this book is your key to solving plumbing problems. As you read these chapters, you will learn about every facet of residential plumbing repairs and replacements. You will even learn how to assess the difficulty level of the job and when you should call in a professional.

The book is broken down into specific sections to make it easier for you to find the advice you need. Whether your drain is stopped up or your faucet won't shut off, this book will tell you how to fix it. In addition to a comprehensive approach to repairs and replacements, you will learn all about plumbing tools and materials.

When repairs are discussed, you will get suggestions on what types of materials to use, as well as finding out about special tools to make the task easier. You will directly benefit from the experience of a seasoned professional. Safety is always of the utmost concern, and this book will show you the safe way to install and repair plumbing. Safety equipment will be described and recommended for certain aspects of your plumbing endeavors.

When you learn to do your own plumbing, you will save money. Plumbers are one of the highest paid trades in the residential field. Hourly rates are commonly in excess of $35, and most companies charge for their travel time. It is not unusual for a simple repair to cost over $100 when you call a professional. With the help of this book, you can put that money to better use.

As you read the following chapters, you will learn to troubleshoot and identify plumbing problems. The book takes you, step by step, through the troubleshooting process. There are many forms to aid you in your search for the problem. By using these professional techniques, you will reduce your time in locating your system's deficiency.

After finding the problem, you will be guided to the effective means of repairing the fault. If repair is not feasible, you will be instructed in the proper procedures for replacing the defective item.

There will be very few areas in your home's plumbing you cannot confidently work with after reading this book. You can also refer to the book as a job manual during your plumbing efforts. With the many illustrations and detailed examples, you will feel competent to work on your own plumbing. Now that you know what this book is about, let's explore how to use it effectively.

How to Use This Book

Plumbing can be very deceptive. Repairs that appear simple can become complicated fast. Jobs seeming nearly impossible can be conquered with minimal effort when you have the proper tools and knowledge. As you read this book, you will gain the valuable knowledge and confidence you need to tackle your own plumbing repairs.

The book is laid out in a logical order to expedite your learning process. The information is complete and written to be understood by the average person. When dealing with complex solutions, the text is simplified so you can easily understand it. When words are not enough, there are pictures and illustrations. It is important for you to follow this advice as it is given. Taking shortcuts can cause big trouble.

Chapters 1 and 2 will get you started. Chapter 1 gives you broad basics on how plumbing works and why it doesn't. Chapter 2 tells you when to call in a professional. Sometimes your best intentions will leave you in the worst condition. Chapter 3 will educate you in the tools of the trade. Chapter 4 discusses the materials approved and most frequently used in residential plumbing applications.

Troubleshooting is the act of defining your plumbing problem. Before you can fix a problem you must know what the problem is. The tips on troubleshooting are very explicit and thorough. In many cases it is simply a matter of trial and error.

Plumbing repairs range from simple to extremely technical. This book covers them all. You will learn to repair nearly every plumbing item in your house by the time you finish this book. The only exception will be in those rare circumstances when the risk is too great for a homeowner to attempt the job. In these cases, you will be told how far you can go before calling for professional help.

There will be times when it is senseless to repair your old plumbing. Either parts will not be available or the cost will be more than is justified. When this happens, you must replace the old plumbing. The sections on replacements will instruct you in how to remove existing plumbing and install new plumbing. Almost every imaginable plumbing endeavor is covered.

As you move through the pages, you will find answers to questions you have not thought to ask. If you are at all handy, this book will make plumbing possible for you.

To maximize the content of this book, you must follow the instructions carefully. Small mistakes can cause big problems when working with plumbing. In some instances, your plumbing duties will involve electrical wiring. Brushing against a hot-wire in your water heater can be fatal. Follow all the safety precautions and don't attempt jobs you don't understand. You could flood your home, electrocute yourself, or ruin expensive plumbing fixtures and equipment. If you follow the instructions closely, you should not have these problems.

Inevitably, something will go wrong at the most inconvenient time. You will need a special part when all the stores are closed, or your new repair will spring a leak hours later. These problems happen even to professionals.

As for unexpected leaks, expect them. Compression fittings work loose. Solder joints can hold for days before blowing apart from stress or vibration. I will provide the best advice possible in avoiding after-the-fact problems, but that risk always exists. The same risks would be present if a professional performed the work. Doing your job by the book will reduce these risks to a minimum.

In seventeen years of working in the trade I have had very few accidents. I have broken two toilets and one tub surround. Considering the volume of work I have performed, these three faults are hardly worth mentioning. Taking your time and working with the proper tools will eliminate much of your risk in damaging parts and fixtures.

You are about to venture into the boots of a plumber. Please remember, the information here is as accurate as I can make it. I have done my best to ensure your success, but if the advice isn't working as described, call a professional. There is no shame in knowing when to call an expert.

1
Basic Plumbing Principles

Plumbing is one area of the home where many do-it-yourself homeowners are intimidated. This fear is not unjustified. A home's plumbing system can be complex and difficult to work with. For the inexperienced weekend plumber, a casual repair can escalate into a full-scale emergency. The simple act of cutting off a faucet can flood an entire house. These emergencies always seem to happen when the stores are closed and the plumbers are charging overtime rates.

Watching a qualified plumber work can make the techniques appear very easy. Seeing a plumber solder copper pipe can make a homeowner feel he or she can do the next repair without help. It is not until you attempt the job that you find there is an art to plumbing. By this time your water is turned off, and the plumbers are all out golfing or fishing. You are left with a leaking pipe and only answering services to call for help.

If you don't like being held hostage by your plumbing, this book can free you. While many plumbing jobs seem easy, most of them require the proper tools and a degree of skill. You can buy the tools, and I can teach you the basic skills. As an experienced master plumber and an instructor for apprenticeship classes, I feel confident in leading you toward a working understanding of your home's plumbing.

This chapter will not directly instruct you on how to master your home's plumbing system. It will give you information on the different types of plumbing in your home and the type of work required to repair or replace it. Later in the book, you will learn on a step-by-step basis how to approach all the plumbing in your house. To get you started, I will discuss each phase of your home's plumbing.

PHASES OF HOME PLUMBING

The Kitchen

Your kitchen can house many types of plumbing fixtures and pipes. The most common elements of a kitchen include a sink and a faucet. In addition to these standard features, the kitchen can hold numerous other plumbing items. You may have a dishwasher or a garbage disposer. Your refrigerator may be equipped with an icemaker.

Disposers and dishwashers fall into the category of *appliance plumbing*. Their internal plumbing is usually repaired by an appliance person, not a plumber. The plumber's job normally stops where the plumbing connects to the appliance. This can include drains and water distribution pipes. The repair and replacement of these items is covered in Chapter 5. All your kitchen plumbing questions will be answered in Chapter 5.

Bathroom Plumbing

This one room contains some of the most crucial plumbing fixtures in your home. There is typically a lavatory, a water closet (toilet), and a tub or shower. In some homes, the bath fixtures can be much more extensive. They may include a whirlpool, bidet, spa, or other specialty fixtures.

Working in your bathroom can call for some flexibility. There is rarely abundant room in and around these fixtures. Working in a vanity, under your lavatory, can make you wish you were a kid again. These cramped conditions call for special tools.

All the tools used to accomplish your plumbing goals are described in Chapter 3. There are many illustrations and instructions for the uses of these tools. Chapter 4 shows you the many different types of plumbing materials and fixtures you may encounter. With these two chapters read and understood, you will be ready to buckle on your tool belt.

Common bathroom fixtures and problems are covered in Chapter 6. You will learn how to work with compression fittings, traps, wax seals, and a host of other commonly encountered challenges. Chapter 6 will make you proficient in bathroom repairs and replacements.

Specialty Plumbing

Chapter 7 delves into the out of the ordinary. This chapter addresses the proper methods for working with less common plumbing fixtures. They include: bidets, whirlpools, spas, and other unusual plumbing fixtures. Whether you want to know how to adjust the air-volume control on your whirlpool or replace the parts on a bidet, Chapter 7 will explain the process.

Getting into Hot Water

Many of us take our hot water for granted until it isn't hot anymore. Chapter 8 describes how to troubleshoot and repair your water heating device. Electric water heaters, gas water heaters, oil-fired water heaters, and domestic coils are all covered in this chapter.

You will learn how heating elements, thermostats, thermo-couplings, and relief valves are replaced. There is important safety information in Chapter 8. Making the wrong moves with your water heater can be fatal. Some aspects of water heater repair and replacement are best left to the professionals. I will tell you how far you should go before calling in the pros.

Waste-Water Pumps

Have you ever been panicked because your basement was filling up with water and there were no plumbers to be found? If your sump pump fails, your basement can flood. Something as simple as freeing a stuck float can solve your problem. When you have a basic knowledge of how your plumbing works, you can fix many problems without help. Chapter 9 delivers the inside information on waste-water pumps.

Filters and Conditioners

Nothing is more frustrating than paying a plumber $40 to change the filter in your in-line water filtration device. To watch the plumber unscrew the housing and replace the filter looks so easy. It only takes ten minutes to complete the job, but you have to pay the plumber for a full hour. Most plumbers will not respond to your call unless you agree to pay a one hour minimum service fee. With a little guidance, you could have changed the filter yourself. Chapter 10 shows you how to do this and much more.

The Water Pipes Inside Your Home

Most people never pay much attention to their water pipes. But when one of these pipes springs a leak, it gets your full attention. Do you know where the valve is that cuts off all the water to your house? If you don't, you had better locate it. When a serious problem develops, knowing where the main cutoff valve is and how to use it can save the day. Chapter 11 takes you on a guided tour of your home's water pipes.

How Important Are the Drains in Your Home?

Everyone knows they are there, but few people understand the function of drains and vents. To the average homeowner, they are just pipes that remove unwanted waste and water from the plumbing fixtures. In simple terms, this is a correct assumption, but there is much more to the drain, waste, and vent (dwv) system. When this system fails, your health is at stake. Chapter 12 shows you how to work with the dwv system and explains potential problems to keep in mind.

What Do You Do When the Drains Won't Drain?

As a working plumber, I have seen homeowners use some unusual methods in attempts to clear blocked pipes. When their crude efforts failed, they were forced to call a professional. In many cases, the work they had done made my job more difficult. The result was aggravation for me and a large bill for the befuddled homeowner. Chapter 13 instructs you in professional techniques to make your drains run smoothly. With this chapter, you may never need to call a plumber again for stopped-up drains.

Underground Plumbing

When you are dealing with underground pipes, you have several obstacles to overcome. First, you must locate the pipes. Then you must assess the proper method to correct the problem. You could dig up your lawn or jackhammer your floor to see the pipe, but you will find out how to troubleshoot the problem without this disruption. If, after evaluation, you determine the pipe must be uncovered, this book shows you how to limit the damage to your lawn or floor. Chapter 14 takes you underground for some revealing facts about your unseen plumbing.

Well Pumps

In rural locations, the well pump is the heart of the plumbing system. When the well pump doesn't work, the rest of the plumbing is helpless. Chapter 15 prepares you to troubleshoot and repair your pump and related equipment. This chapter takes you from the bottom of the well to the main water distribution pipe in your home. Everything in between is discussed and described. You will learn about controls, pressure tanks, valves, gauges, wiring, and other related equipment.

Where the Grass is Always Greener

Your septic system can go unnoticed for years, but when it malfunctions, you will notice it. There are some preventive maintenance ideas you can employ to reduce the risk of failure in the septic system.

If the system does fail, Chapter 16 guides you through the steps of evaluating the failure. There are tips on how to find your septic tank and when to have it pumped out. In addition, you will learn how your system works and what you can do to repair it.

All the Little Things

Chapter 17 takes on all the small parts, valves, and miscellaneous plumbing items in your home. It provides detailed instructions on what these devices do and how you can repair or replace them.

Thinking of Adding a Bathroom?

When your remodeling plans involve plumbing, you can be in for an expensive lesson. Plumbers love remodeling work. They know this is where they can make some of the best money for their time spent on the job. Remodeling requires working with old plumbing and unknown conditions. How would you connect plastic pipe to cast iron pipe? Do you know how many fixtures your existing plumbing is capable of serving? If you add another bathroom, you may need to upgrade the existing pipes to a larger size.

If you call in a professional, will you know if what is suggested is an accurate assessment of your situation? Chapter 18 is dedicated to remodeling and adding to existing plumbing. Plumbing for large jobs may be more than you want to tackle, but the information here will help you keep the plumbers honest. If they tell you your $3/4$-inch water line is not adequate, you can check the charts in this book. These charts show you how to estimate the size of pipe needed for your application. Chapter 18 will provide enough information for you to talk intelligently with the professionals you hire to do the job.

Putting on the Engineer's Hat

Chapter 19 talks about sizing and designing new plumbing systems. This information can be used in planning your new plumbing system. Using this information, you can develop a desirable plan to show your contractor. I will give you suggestions on fixture locations and how to describe your desires to the plumbing contractor.

Figure 1-1.

Figure 1-2.

Figure 1-3.

Figure 1-4.

Figure 1-1. *Always use safety glasses.* ***Figure 1-2.*** *Cleaning a copper fitting with a brush.* ***Figure 1-3.*** *Cleaning a fitting with sandpaper and a pencil.* ***Figure 1-4.*** *The wrong way to sand pipe. Model: Andrew E. Wallace.*

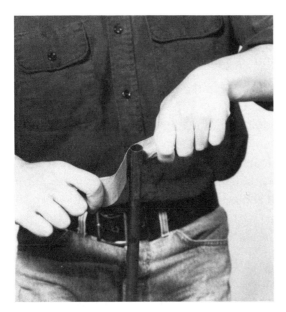

Figure 1-5.

Figure 1-6.

Figure 1-7.

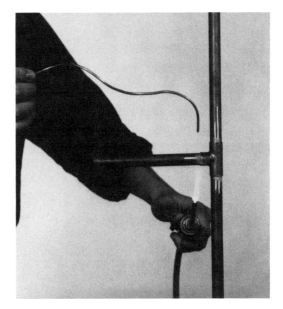

Figure 1-8.

Figure 1-5. The right way to sand pipe. *Figure 1-6.* Applying flux with flux brush. *Figure 1-7.* Torch placement for a vertical job. *Figure 1-8.* Torch placement for a horizontal job. Model: Andrew E. Wallace.

Picking a Pro

There will be times when you need a professional. When this time comes, how will you know who to call? Will you go to the phone directory and call the first advertiser you come to? Not all plumbers are the same. Some specialize in service work; others concentrate on commercial work. Chapter 20 shows you how to pick your pro.

An Ounce of Prevention

It may be a cliché, but it's true. Routine maintenance can save you from major repair bills. In Chapter 21, you will discover the most effective ways to preserve your plumbing.

BASIC RESIDENTIAL PLUMBING TECHNIQUES

As you contemplate becoming a weekend plumber, what do you believe is the most difficult task to accomplish? If you think soldering copper pipe is the hardest, you are among the majority. For the average person, soldering a pipe joint so that it won't leak seems to be an insurmountable task.

Soldering

Soldering requires some practice, but it is not very difficult when you follow the proper steps. If you clean the pipe and fittings, flux them, and master the torch, you will have no trouble. A hand-held propane torch is all you need for residential jobs. Let's examine how soldering is done in detail.

This description is based on normal conditions when soldering copper pipe. If the pipe you are soldering has water in it, you will need to make a few procedural adjustments. The second example covers how to deal with water.

Eye protection should be worn when soldering (Figure 1-1). Long sleeves and gloves provide additional protection from hot, dripping solder. The first step is to clean the pipe and fittings (Figure 1-2). The easiest way to clean the fittings is with a special wire brush. These fitting brushes are sold very

inexpensively in different sizes to fit most fittings. You can also use sandpaper to clean the fittings (Figure 1-3). On small diameter fittings, wrap the sandpaper around a pencil to make it easier to get into the fitting. Turn the fitting brush clockwise until the inside of the fitting shines.

Clean the end of your pipe with a light-grit sandpaper. Copper pipe can have jagged, sharp edges where it has been cut (Figure 1-4). Never cup your hand around the pipe and rotate the sandpaper to clean it. Instead, hold the sandpaper at each end and work it across the pipe as if you were shining a pair of shoes (Figure 1-5). This method will protect your hands from any sharp edges. Sand the pipe until it shines and is free of dirt.

Use a flux brush to apply flux to the inside of the fitting and to the end of the pipe (Figure 1-6). It is not necessary to create a thick layer, but be sure to have flux on all areas to be soldered. Flux is an eye irritant; do not rub your eyes when you have flux on your hands. After applying the flux, insert the pipe into the fitting. With the pipe seated fully into the fitting, you are ready to solder the joint.

Using a 95/5 or no-lead solder, pull a string of the solder off the roll. Using a striker or match, light your torch. Hold the flame under or beside the fitting as illustrated. For small pipes like those found in most homes, the heating time will not be long. You will see the flux melt as the copper becomes hot. The flux is cleaning the pipe and will help the solder run around the joint.

When the flux melts, start testing the joint with the solder. Place the solder at the lip of the fitting. When the temperature is right, the solder will run around the fitting. If the pipe is dirty or too cold, the solder will bead or clump up. If the joint is too hot, the solder may not form a good joint. You may have to make several practice attempts before learning how to gauge the proper temperature for making the solder joint (Figures 1-7 and 1-8). When soldering, hold the heat source as illustrated for vertical and horizontal jobs. The heat will pull the solder into the fitting for a good joint.

Soldering with Water Present

In making repairs, it is not unusual for the pipe to be holding residual water. This water can make your soldering efforts very frustrating. When the joint is heated, the water will turn to steam and blow holes in the solder. These holes will cause your joint to leak. This is one of the most common problems homeowners run into and give up on.

There are two simple ways to work around this problem under average circumstances. You can solder a valve onto the pipe before joining the two sections. Open the valve and solder it on one end of the pipe. If you do not open the valve before heating the body, the washer in the valve may melt. Do not expose your body to the open end of the valve, as steaming hot water may spurt out the end of the valve. By having the valve open, the steam can escape without affecting the solder joint. When the valve cools, close it and join your two pipes. Now you can solder the pipe and fittings without the water coming past the valve. Valves are typically made of thicker material than copper, and they therefore take longer to heat to the proper soldering temperature.

Weep-hole fittings are another simple solution (Figures 1-9 and 1-10). They are available as couplings and elbows. By using these fittings you can unscrew the weep-hole cap. The weep-hole will allow the steam to escape while you solder the joint. When the joint is made and cool, screw the weep-hole cap back on and you are in business.

When you can lay the solder on the fitting and see it run around the joint, you probably have a good seal. Do not jar or move the joint until it cools. Any movement while the joint is hot could cause a void in the seal and consequently a leak. When the joint cools, you can wipe it with a cloth to remove excess flux. If you do not wipe the joint, the flux will probably turn green on the outside of the pipe.

PLASTIC PIPE PRINCIPLES

Gluing plastic pipe together probably seems so easy you could not make a mistake. Don't you believe it! There are many different types of plastic pipe, and each of them requires proper preparation and solvent application. You are going to learn to work with ABS, PVC, CPVC, PB, and PE plastic pipe. Some of these plastics are used for drinking water and others are used for drains. Depending on the type of pipe, you will use glue, crimp rings, or stainless clamps to complete your pipe joint.

Plastic Drain Connections

When working with plastic pipe for drains, you have several steps to follow in making a connection. Clean any burrs from the end of the pipe. Clean the end of the pipe and the fitting with a cleaner recommended for the type of pipe you are using. Next, apply a primer coat to the end of the pipe and fitting with an approved primer. Apply a solvent cement of the proper type to the pipe and fitting. Insert the pipe into the fitting until it is seated fully. Turn the pipe a quarter of a turn to distribute the glue. Allow the joint to bond before moving the pipe.

Plastic Water Connections

With CPVC, the connection procedure is the same as the above instructions for drain pipes. For PB

Figure 1-9. Weep coupling.

Figure 1-10. Weep elbow.

pipe, you need special equipment. You need a crimping tool, crimp rings, and a crimp measurement device. Follow the manufacturer's instructions for installing PB pipe. Basically, you place a crimp ring on the pipe and insert a male-insert fitting into the pipe. Position the ring over the insert portion of the fitting. Crimp the ring with the crimping tool. Use the measuring device to check the quality of the crimp.

When working with PE pipe, you will use insert fittings and stainless steel clamps. Slide your clamps over the pipe and insert the male-insert fitting into the pipe. Position the clamp over the insert portion and tighten the screw on the clamp. It is best to use two clamps on each connection. (Note: Additional information on these procedures is included throughout the book at appropriate points.)

THE CHINA CHALLENGE

You will need to develop a gentle touch when working with china fixtures. Tightening the bolts too much on your new toilet can leave you with broken china. There is a fine line between too tight and not tight enough. With the proper tools and instructions, you can learn to master the challenge of china.

SHOCKING EXPERIENCES

Many plumbing fixtures depend on electrical current to make them operational. When you are working on these fixtures, you must use extreme caution. Your water heater can deliver a 220-volt jolt without warning. This is enough juice to melt your screwdriver and cause serious damage to your body. This risk is amplified when you are standing in water. If your water heater needs to be replaced, there is a good chance of water on the floor. Before attempting any repairs or replacements, you must obey the safety rules.

DRAIN CLEANING

Big-league drain cleaners can be rented. When you rent these devices, the merchant may explain how the machine works. It is unlikely that he will warn you of the problems that can arise from the use of these powerful drain cleaners. It is possible to break your pipes or have the cable become stuck in the pipe, or you could lose a finger.

If you are using electric drain cleaners, you need to know the proper methods to employ. The torque from these machines can literally rip and tear at your body. I don't mean to be gruesome, but you must respect the power these blockage-busters have. You run the highest risk when you are cutting tree roots out of your underground pipes with these machines.

TOOLS

As a weekend plumber, you will work with a variety of hand tools. Many of these are common tools you may already have. A few of the tools are specialized plumbing tools. Most of the specialized tools can be rented if you don't want to invest in their purchase. Again, safety is your first concern.

PREPARING TO MOVE AHEAD

For basic plumbing repairs and replacements, you don't have to learn hundreds of different principles. You will use many of the same principles for multiple tasks. Cutting pipe is relatively easy. Soldering can be learned with a little practice. Making solvent and clamp joints on plastic pipe is simple when you follow the instructions. The remainder of what you must learn is equally obtainable. This book is very comprehensive in its approach. You should be able to determine whether the job is more than you can handle *before* you have done damage to yourself or your plumbing.

Now that you have a feel for the basics, it is time to move ahead. In Chapter 2, you will learn about doing more harm than good. I know you are anxious to jump right to the chapter you need for the task at hand, but you will do well to read Chapter 2 first. Before picking up your pipe wrench, turn to the next chapter and learn from other people's mistakes.

2
Doing More Harm than Good

There are many occasions when homeowners try to save money and wind up creating very costly mistakes. They assume they have the ability to repair their own plumbing and go blindly about it. This recklessness often results in more trouble than they had to begin with. This chapter will educate you from the mistakes of others. In my years as a plumber, I have seen many instances where the homeowner didn't know when to call for professional help. He ventured into the unknown and by the time he called me, the problem was much worse than when he started the work.

One too many turns of the wrench can send water flying through your house. Trying to replace your own garbage disposer can incapacitate your kitchen sink until you find a plumber. Cross-threading a fitting can require replacing an entire plumbing fixture. This chapter is written to help you avoid these types of problems.

OUTSIDE THE HOUSE

Most people think of the inside of their homes when they conjure up ideas of plumbing problems. While it is true that most plumbing problems occur under your roof, many of them can be found outside. Here are some examples of situations when the homeowner did more harm than good.

The Lost Well Pump

Harry was having problems with his water pump. It was not running properly, and sediment was coming through his faucets. After talking to his neighbor, Harry felt he could fix the pump. His neighbor had recently experienced similar problems and had called a plumber. The plumber came out and fixed the problem in less than an hour. Harry's friend told him how the plumber had fixed the pump.

Harry went to his well casing and removed the cover. Looking down the well he saw a large black plastic pipe descending into the well. Following his neighbor's instructions, Harry got a grip on the pipe and pulled it up and shook it. He repeated this process several times and then went inside to see if his home repair had worked. According to his neighbor, all the plumber had to do was shake the pump and the problem cleared up.

Harry turned on the faucet, but the problem still existed. He went back to the well with his son. The two of them heaved and shook the pipe. All of a sudden, there was a jerk and then little resistance. The pipe was much lighter now, and Harry could only assume what had happened. He went back to the house and turned on the faucet. There was only a trickle of water and then there was none. His pump had plummeted to the bottom of the well.

All the stress placed on the plastic pipe and nylon fittings caused the connection between the pipe and pump to sever. When my plumbers arrived, there was nothing to be done but for Harry to buy a new pump. If he had called a professional first, he would have had to pay less than $100 for the repair. With his efforts, the job cost over $700.

Harry told my plumbers about his neighbor's advice and wanted to know what went wrong. There could have been many reasons for his pump disaster. His pump was hung with nylon fittings; maybe his neighbor's was hung with brass fittings. Harry didn't have a rope tied to the pump securing it to the casing; maybe the pump next door did. The exact reason will never be known, but Harry had to buy a new pump due to his repair attempt. He went about the job in the wrong way and without any knowledge of what he was doing. Acting on the advice of a well-meaning neighbor cost Harry hundreds of dollars.

The Hard-Nosed Hose Bibb

Larry's outside hose bibb had developed a steady drip. Larry went to the local hardware store and purchased an assortment of faucet washers. He had read something about changing washers and was sure he could handle the job. It was late on a Saturday afternoon when he began his work.

There was a valve in his garage that controlled the water flowing to the hose bibb. Larry turned this valve off and went outside to remove the stem from the hose bibb. After fussing with the retaining nut, Larry finally had the stem coming out of the hose bibb. In a flash, Larry was soaking wet. Water was blowing out of the end of the open hose bibb at full pressure.

Larry ran inside the garage to see if he had turned the valve the wrong way when cutting the water off. The valve seemed to be closed, but water continued to flood the yard. Grabbing a pair of pliers Larry cranked down on the valve to close it. He checked and the water was slowing down. Assuming the valve must just be stuck, he turned it more with the pliers. With his last turn, the valve started spraying

water *into* the garage. Now he had water running full-bore outside and spraying inside.

Panicked, Larry couldn't remember where the main shut-off for his house's water supply was. He looked but couldn't find it. In the meantime, Larry's wife was frantically calling plumbers. When she reached my company, I dispatched a plumber immediately.

My plumber arrived and found the garage partially flooded and the lawn heavily eroded from the water pressure. After cutting off the water, my plumber replaced the broken valve and repaired the hose bibb. For an experienced plumber this was a routine call.

If Larry had called us to fix the leaking hose bibb during normal business hours, it would have cost him $33. Having a plumber come out on a Saturday meant paying time-and-a-half. In addition, the plumber had to replace the valve Larry broke. This might not have been necessary if the plumber had been called to do the work in the first place.

Larry's plumbing bill for this call totaled over $100. By trying to fix the problem himself, Larry incurred many extra costs and much lost time. In addition to the extra plumbing fees, he had to clean the water out of the garage and repair the damage done to his lawn. This was the result of inexperience and a lack of knowledge.

The Summary

Both of these hapless homeowners made common mistakes. Harry acted on the advice of a neighbor and didn't know when to stop. Larry tried to recall instructions he had read some time ago to repair his hose bibb. In both cases, the homeowners did more harm than good.

Neither of these jobs required a master plumber. If either of these men had owned this book, they would not have caused so much trouble. While you do not have to hold a master plumber's license to do household repairs, you must have a clear understanding of the job you are taking on. In both instances, the men knew just enough to be dangerous. Before you try your hand at plumbing, be sure you are competent to do the job. This book will tell you everything you need to know in order to avoid these common mistakes.

INSIDE THE HOUSE

A Drain Cleaning Disaster

When Sue's daughter flushed the toilet, water came up instead of going down. Instead of calling a plumber, Sue tried using a plunger on the toilet. This attempt made no progress, and Sue called her husband. Harold said he would take care of it when he got home. That evening Harold flushed the toilet to see if it had fixed itself. It hadn't, and water flooded the bathroom floor for the second time.

Harold went to the basement and found his flat-tape sewer rod. He thought of removing the toilet, but decided to use the clean-out plug in the basement instead. The clean-out plug was in the main drain, about chest high off the basement floor. Without testing any other plumbing fixtures, Harold started removing the screw-in plug from the clean-out.

When the plug became loose, it blew out of the pipe, hitting Harold in the chest. Following the plug was all the backed up sewage held in the pipe. Needless to say, Harold and the basement were a mess. Since he was already covered with undesirable substances, Harold proceeded with his snaking of the drain. After putting fifty feet of snake into the pipe, he still had not hit a clog.

Harold replaced the plug and had Sue flush the toilet. Sue told him it worked fine. She tried the toilet a second time and again it worked admirably. Assuming he had corrected the problem, Harold cleaned up the mess and took a shower. Later in the night, the toilet backed up again. Perplexed, Harold called a plumber.

Harold's problem turned out to be not a blocked drain but a full septic tank. When the toilet had been tested, it worked because the volume of liquid in the drain had been released on the basement floor. Being a 4-inch pipe, it took a considerable amount of water to refill it before the toilet backed up again.

Harold had paid a high personal price to learn a valuable lesson. If you have a septic tank and your drains back up, check the level in the septic tank before calling a plumber. It cost Harold $49.50 to be told he needed to have his septic tank pumped to solve the problem.

If Harold had pulled the toilet and snaked the drain from that point, he would not have received his unpleasant drenching. He would also have been able to observe the water level in the pipe as he ran the snake down the drain. There would have been no doubt that he had not cleared a blockage and eliminated the problem.

Code Complications

Bob decided to add a bathroom in his basement. Bob was very handy and a quick learner. He read several books and was able to install a complete bathroom in his basement successfully and without the help of professionals. Bob was very pleased with his installation.

He had broken up the concrete floor and connected the new plumbing to the existing drains beneath the floor. After running his water pipe in the ceiling joists, he drywalled the ceiling. Bob was able to add the bathroom and a family room to his basement without any outside help.

During the first year of use, Bob enjoyed the new space he had created. The additional plumbing made the basement much more comfortable and usable. When the first of the new year rolled around, Bob was visited by the tax assessor. As the assessor performed her inspection of the property, Bob bragged of his prowess as a carpenter and plumber. He showed off his handiwork with the utmost pride.

Two days later, Bob answered the door and was confronted by a code enforcement officer. After identifying himself, the code officer asked to inspect Bob's basement. To jump to the end of the story, Bob had never been issued the proper permits for the work he did in the basement. As a homeowner, he could have obtained the necessary permits, but he never bothered to apply for them.

In this particular case, the code officer was very lenient. He had Bob swear to the method of installation of the plumbing and let it go with only a cash fine. He could have required Bob to tear down the ceiling to expose the water pipe installation. By code, he could have made Bob open the concrete floor to expose the drainage work.

Bob had done all the physical work perfectly, but he made a major error in not obtaining a permit for the work. This type of mistake could be extremely expensive if you run into a code officer with less appreciation for your circumstances. Before installing new plumbing, always inquire about the need for permits. Most new installations require a permit and inspection.

OUTSIDE PLUMBING POINTERS

The plumbing found outside your home may require different techniques than inside plumbing. You may become involved in excavation, underground wiring, and a number of other unexpected obstacles. Working on your well pump cannot be compared to troubleshooting your kitchen faucet. Some of the same basic principles apply, but the methods for repair or replacement may vary a great deal.

In the two examples for outside plumbing, the homeowners made common mistakes. They listened to advice from inexperienced people and they didn't use common sense. As you are learning plumbing, common sense plays a primary role in your success. If the people in the examples had used more logic and planning, they might not have had such bad experiences.

Plumbing often creates unexpected problems. Since you don't have a stocked plumbing van in your driveway, you must anticipate these problems. If something doesn't seem to be going right, don't force it. Needing repair parts late on a Sunday night can leave you without water. As you read the various sections of this book, you will be advised of extra parts to have on hand when attempting a repair or replacement. This advice is based on years of experience in needing something I didn't have.

INSIDE PLUMBING

Most plumbing problems occur inside your home. The nature of these problems can range from a dripping faucet to a ruptured water pipe. The predicament can cause major flooding in your home or a constant grating on your nerves. In either case, you will encounter most of your plumbing challenges within the home.

This book covers all the major categories of plumbing within residential property. You will learn to cut cast iron pipe, solder copper pipe, and set china fixtures. As thorough as the instructions are, there will be times when the job doesn't go as planned. By using the book as a manual and pre-planning your project, you can avoid most of the mistakes commonly made in handyman plumbing.

What follows in this chapter are some examples of what *not* to do. These examples illustrate how a simple task can develop into a major project. They go on to show how easy it is to make costly errors. As you read the examples, you may see that they strike close to home. Many of these incidents are constantly recurring among individuals attempting to do their own plumbing.

The Broken Toilet

Denise decided to replace her old toilet with a new low-flush toilet. She liked the idea of conserving water and money. After pricing the job with many plumbing firms, Denise came to the conclusion that she could do the job herself. She had picked up a pamphlet on plumbing at her local building supply store and felt she was capable of doing the job.

Denise bought her new toilet combination and installation supplies. She was ready to make the swap and had very little problem removing the old toilet. When she removed the old toilet, she noticed her piping arrangement didn't match the one in the pamphlet. The illustration showed a ring around the pipe with slots to accept the closet bolts. Her toilet's drain pipe didn't have a flange. The old toilet had been attached to the floor with lag bolts.

This discrepancy confused Denise, but she didn't give up. She called the store where she purchased the toilet and received free advice. She found that in the old days, toilets were set right over the pipe and screwed to the floor. This method is still possible with new toilets, but it is best to install a flange. Denise elected to install the new toilet in the same manner as the old one.

She placed her wax ring over the pipe and set the toilet bowl on it. After carefully aligning the bowl, she screwed it to the floor. Next, she began to install the tank on the bowl. During this process, she tightened the tank-to-bowl bolts too tight. The result was a cracked toilet tank. Angry and discouraged, Denise took her cracked tank back to the store.

She was informed that the tank was not defective and the crack was a result of her efforts. Under these conditions, the store would not exchange the tank. Denise bought a new tank and went home to install it. This time, she was very careful not to overtighten the bolts. After making the water connection to the tank, she turned the valve on and the tank filled with water.

Water leaked around the tank-to-bowl bolts. She tightened the bolts until the drip stopped and everything seemed fine. She flushed the toilet several times and felt she had made a good installation. Feeling proud, she went downstairs to put her tools away. Walking through the formal dining room, she saw water running out of her ceiling light fixture. Panicked, she raced upstairs to the bathroom.

There was no sign of a leak in the bathroom, but water continued to drip out of the light fixture below. Ultimately, Denise called us to fix her problem. The water in the light fixture was coming from the toilet's drain. The wax seal Denise used did not seat evenly, and when the toilet was flushed, water escaped around the pipe. The water collected in the ceiling and ran out the light fixture.

In Denise's case, her efforts cost more than she saved. Having to buy an additional tank was the beginning of her financial loss. Making a poor connection between the toilet and the drain caused her to call in professionals. In the end, she would

have been much better off to have called us in the first place.

Denise's biggest mistake was working with limited information. She had the basic skills to do the job, but was not well-informed for the project. Breaking the china could have happened to anyone, but it didn't have to. When she installed the second tank, she didn't break it. By making the bolts snug and filling the tank with water, she was able to tighten them just enough to make a seal without breaking the tank. If she had read about this procedure before attempting it, she would not have broken the first tank.

With a comprehensive manual, she would have figured out where the water was coming from through the light fixture. This information would have saved the cost of a professional service call. The key to success is adequate information and a clear understanding of it.

Here's Rust in Your Eye

Frank was a competent handyman. He had worked with his hands for years and could grasp new procedures easily. After reading about faucet replacement, Frank decided to replace his kitchen faucet. The old faucet had been on the sink for years and was heavily rusted at the mounting nuts.

After assuming a position only a plumber or a contortionist could achieve, Frank began his work. The conditions under the sink were restrictive, and the mounting nuts were seized by rust. While attempting to turn the nuts, one of them shattered. The debris and rust from the nut fell into Frank's eye. The injury required medical attention, and we had to complete the plumbing job for Frank.

Frank's injury would have been avoided if he had been wearing safety glasses. In Frank's case, his negligence caused him pain and financial loss. It could have been worse — it could have cost him his eyesight. When you are using the information in this book, don't disregard the safety recommendations. They are important and can save you money, time, and pain.

Playing with Fire

In cold climates, it is not unusual for plumbing in the outside walls of a home to freeze. Many times, these frozen pipes can split and flood the home. Most of the pipes are copper and require repair by compression fittings or soldering. The standard procedure is to solder them. This requires the use of an open flame from a torch. Outside walls are typically filled with insulation. Many of the components of the wall are flammable.

A plumber I knew nearly burned down a home by making a stupid mistake. He was repairing a broken water pipe in an outside wall. As he was soldering the pipe, the insulation caught on fire. Fortunately, his boss reacted quickly by opening the wall and extinguishing the fire. The potential for fire damage is always present in these conditions, and it escalates when the job is being done by a homeowner.

This book will remind you of potentially dangerous situations you may wish to avoid. It is impossible to cover every conceivable outcome to a job in a book. However, I believe this is the most complete training manual for homeowners available on plumbing repairs and replacements. If you temper the text with logic and common sense, you should enjoy successful outcomes to your plumbing ventures.

The previous examples are included to give you an idea of the type of risks present in residential plumbing. You risk personal injury, property damage, and financial losses. Even with these risks, countless numbers of people attempt plumbing without any preparation. You have purchased this book, which proves you are ahead of the crowd. With the aid of this information, you will be much better prepared to deal with unexpected situations.

BEYOND THE PIPE WRENCH

In today's society, there is much more to plumbing than a box of tools and a strong back. Plumbing fixtures are much more complicated than they were fifteen years ago. Now, you can have a faucet that comes on automatically when you place your hands under it. There are complex whirlpool tubs and intricate product designs to baffle you. Code re-quirements continue to become more stringent. Even as a weekend plumber, you must show respect for the laws and plumbing codes.

In most cases, a homeowner can obtain a permit to perform plumbing on his residence. This rule may vary in different jurisdictions, but it tends to hold true in most locations. Replacing a water heater requires a permit. Many professional plumbers scoff at this rule and ignore it. They are taking a huge risk. Water heaters can become explosive missiles under certain conditions.

If a water heater is installed without a relief valve, it could blow up, destroying your home or killing you and your family. This is not a minor code infraction, it is a potential time-bomb. This type of information is included in the technical chapters of this book. It is wise to consult with your plumbing inspector before doing *any* significant work. Replacing a faucet or snaking a drain won't require a permit, but replacing fixtures may.

Many of the code requirements may seem trivial, but they exist for a reason. Normally, the code is efficient in protecting your health and safety. Since codes change and some areas use different plumbing codes, you should confirm your plans with the code enforcement office prior to installing new plumbing.

Chapter 3 educates you in the most common tools of the trade. In many cases, you can substitute tools and still accomplish your goal. Chapter 3 shows the best tool for each job. In the technical chapters, I will cross-reference many tools you can use in place of specialized plumbing tools. There are a few occasions when only the described tool will do the job effectively. The book will advise you of these times and make it clear what tool to use.

After Chapter 3, you will learn about approved materials in Chapter 4. Once you have completed Chapters 3 and 4, you will be ready to work. Chapter 5 begins the technical chapters. Each chapter is named for quick reference as you encounter your plumbing problems.

Now, let's see what you need to fill your toolbox.

3
Tools of the Trade

Tools are an essential part of the plumbing business. With the proper tools, even the most difficult job becomes easier. Since you are not a professional, you will probably not have many of the tools designed specifically for plumbing. When you find a tool you need, you can purchase it or, in many cases, rent it.

Most communities support at least one rental center. These rent-what-you-need stores can be a lifesaver. Many of the tools used by professional plumbers cost several hundred dollars. Some of them may cost several thousand dollars. This is a big expense for a professional and an excessive cost for the consumer. When you need a large drain cleaner, you can rent it for between $50 and $100. To buy this piece of equipment could easily drain $2,500 from your bank account.

With an increased demand by consumers and professionals, rental centers stock a huge assortment of plumbing tools and equipment. Many professionals choose the rental route. A plumber may not need a jackhammer frequently enough to justify its purchase. However, when he needs a jackhammer, he can rent it. This is good business and a sound approach to solving occasional problems.

For you, there are many tools you would do well to rent. The sewer machine is only one example. You may wish to rent a cast iron pipe cutter. These dandies sell for several hundred dollars but can be rented for less than $50. This chapter is going to

expose all the tools you are likely to work with in your plumbing tasks. When the time comes to acquire these tools, seriously consider renting them.

PIPE CUTTERS

Pipe Cutters for Copper Pipe
You will need a pipe cutter at some time in your plumbing endeavors. The type of cutter you use will depend heavily on the type of pipe you are cutting and the conditions under which you are cutting it. For average conditions, you will use your pipe cutter for copper pipe. Copper pipe can be cut with a hacksaw, but it cuts much more easily with the proper cutter (Figure 3-1). A standard copper cutter will handle pipe from 1/8-inch to 1 1/4-inch diameter. This same style cutter is available for pipe up to 6 inches in diameter.

There are times when space limitations will not allow the use of this type of cutter. In tight conditions, you can use a mini-cutter. These miniature cutters will perform well in extremely confined conditions (Figure 3-2). If space does not allow the use of these tiny cutters, you will have to use a hacksaw blade. You can use the blade by itself or in a special handle. These special hacksaw devices are very effective in hard-to-reach places (Figure 3-3).

Reciprocating saws are very effective in cutting copper in close quarters (Figure 3-4). This type of saw leaves a rough edge but makes the job much easier in difficult situations. However, if the copper pipe is not secure, this saw will vibrate the pipe and could break existing solder joints.

As a professional plumber, I have worked with all these cutters. The hacksaw is my last choice and the roller-cutter is my favorite. In tight spaces, I prefer the mini-cutters.

Pipe Cutters for Plastic Pipe

Plastic pipe can be cut with a variety of devices, depending on the type of plastic (Figure 3-5). A hacksaw is very effective for the average home-owner's use in cutting all types of plastic. Many plumbers buy a special saw to cut plastic pipe. The saw has an offset handle and can handle many types of materials (Figure 3-6). In addition to these two types of saws, there are times when you need something different.

In close quarters, a strong blade with a handle is the perfect tool. This type of design allows for maximum use in minimum spaces (Figure 3-7). Cutters like those used for copper are popular with some professionals. All cutters of this type are equipped to accept replacement cutting wheels (Figure 3-8). When the wheels are worn, or you wish to cut a different type of pipe, you simply change the cutting wheel. This type of cutter combines easy cutting with clean cuts.

For small plastic pipe, you can use a different style cutter. These clipper-type cutters are effective on plastic pipe up to about 2 inches in diameter (Figure 3-9). Once in a while, you may need to cut your plastic pipe from the inside; for example, in the case of a pipe for a toilet. To cut pipe from the inside, you need a special cutter. These inside cutters have cutting wheels that expand against the pipe. As you tighten and turn the tool, it cuts the pipe (Figure 3-10).

After seventeen years of cutting plastic, I still use a hacksaw for most of my work. If I am cutting pipe larger than 4 inches in diameter, I use a power saw or a regular handsaw. The handsaw is the type used by carpenters to cut wood. The power saw is usually a reciprocating saw. For most applications, I prefer a quality hacksaw.

Pipe Cutters for Cast Iron Pipe

There are special blades available for reciprocating saws to cut cast iron, but this method takes a long time. You will also burn up your blades quickly. For cutting cast iron, I use a ratchet-style snap-cutter (Figure 3-11). These tools are usually available at rental centers, and they make cutting cast iron a snap. You lay the chain around the pipe, tighten the knob, and crank the handle. That's all there is to it. In moments, you have cut the pipe and are ready to move on. With saw blades, you could spend twenty minutes trying to cut the same pipe.

If you have plenty of room, you may want to use a roller-cutter. These heavy-duty roller-cutters are fast and efficient but require enough room to turn them around the pipe (Figure 3-12). The snap-cutters are most often your best choice when cutting cast iron pipe.

Cutters for Steel Pipe

The cutters designed for steel pipe work on the same principle as the roller-cutters for copper. Steel cutters are heavier and have long handles to give you leverage when cutting (Figure 3-13). Some of the cutters are equipped with two handles to make cutting large pipe easier (Figure 3-14).

You can cut steel pipe with a hacksaw or a reciprocating saw. With a fine-toothed blade, either of the saws is adequate on small pipe. Since steel pipe is rarely used in today's residential applications, I am not going to expand on all the possibilities. The primary use of steel pipe in today's home is for the installation of gas pipe. Working with gas should be left to the professionals. Because of the hazards of a homeowner working with gas, this book will not go into all the tools available for that type of work.

DRAIN CLEANING TOOLS

There are numerous tools available for clearing stoppages in plumbing. When your toilet doesn't flush properly, the first tool to try is a closet auger. The auger is placed into the toilet bowl and turned as it goes through the trap and down the drain (Figure 3-15). Closet augers are very effective in clearing stoppages within a few feet of the toilet.

For the homeowner, flat-tape snakes are a viable choice when drains won't drain (Figure 3-16). These snakes are easy to use and are safe for the consumer. In some types and sizes of pipes, these rigid snakes cannot negotiate the turns in the drain. When this happens, you can try a spring snake. For small drains, a hand-operated spring snake can do the job (Figure 3-17). These are safe and easy to use.

When you have a stubborn stoppage, you will need an electric drain cleaner. These units are expensive and potentially dangerous. (For proper operation refer to Chapter 13.) These power snakes can be rented by the hour or by the day. They come in many sizes to meet all your drain cleaning needs. For sink drains, you can use hand-held or brace-mounted units (Figures 3-18 and 3-19).

For larger drains, you will want to use a bigger machine. These machines can clear blockages and cut roots. Many of them are equipped to be foot-operated (Figure 3-20). These large machines come in many designs and sizes. As you can tell by the illustrations, there is a drain machine for every need (Figure 3-21).

In addition to these drain openers, there are several other specialty items available. The one most worth mentioning is an expandable water bag (Figure 3-22). This device attaches to a garden hose and works with water pressure. The use of this unit is described in Chapter 13.

POWER DRILLS

Right-angle drills are a must for professional plumbers. They allow the plumber to drill holes where it would be impossible with other types of drills (Figure 3-23). These drills are rarely needed by the average homeowner. A standard pistol-type drill will serve most of your needs (Figure 3-24). An assortment of drill bits can be collected as you need them. For large holes, a hole saw kit is a good investment for the part-time plumber (Figure 3-25).

TORCHES

Full-time plumbers depend on one of two types of torches. They are air-acetylene torches (Figure 3-26) and propane torches (Figure 3-27). The propane torch uses the same gas as the hand-held torches most homeowners choose. A professional torch outfit costs over a hundred dollars. For the repairs around your home, a simple hand-held propane torch is all you should need.

LADDERS

Plumbing often requires the use of a ladder. Normally, a stepladder is all you will need (Figure 3-28). Occasionally, you may need an extension ladder (Figure 3-29). When you invest in a ladder, buy a good one. Being at the top of a cheap ladder when it collapses is no fun.

HAND TOOLS

Hand tools are the heart of your toolbox. Without hand tools, you won't get much plumbing done. This section on hand tools covers a broad range of equipment.

Shovels
Shovels are one of the first tools a plumber's helper learns to use. A good round-point shovel is the best choice for most plumbing jobs (Figure 3-30).

Hammers
A plumber has a need for small hammers and sledgehammers. Any type of these hammers will get the job done. I prefer a 20-ounce straight-claw hammer (Figure 3-31) and an 8-pound sledgehammer (Figure 3-32).

Pliers
Pliers play a large part in the day-to-day work of a plumber. The most frequently used pliers are water-pump pliers or tongue-and-groove (t&g) pliers. These angled pliers are ideal for many plumbing applications (Figure 3-33). Needle-nose pliers are useful for reaching awkward items (Figure 3-34). Lineman's pliers (Figure 3-35) come in handy when working with electrical wiring. Plumbers frequently work with wiring at water heaters, dishwashers, disposers, and related equipment. Straight slip-

joint pliers can be helpful but are not regularly used by the pros (Figure 3-36).

The potential number of different pliers available boggles the mind, but these are the ones used most often by working plumbers. The tongue-and-groove pliers are by far the most frequently used pliers in a plumber's toolbox.

Snips

Almost every plumber's box will have at least one set of snips in it. Many plumbers carry three sets of snips. These snips are designed to cut in three different directions. They cut left, right, and straight (Figure 3-37). Snips are used for many jobs. They are used to cut sheet metal, hanging strap for pipes, and almost any other task you could imagine. If you only buy one set, buy the set that cuts straight. If you will be doing much sheet metal work, invest in the complete set.

Chisels

Every good plumber carries chisels in the toolbox. They include wood chisels and cold chisels (Figure 3-38). The chisels are needed to open up areas to make plumbing installations and to gain access to plumbing.

Measuring Devices

A retractable tape measure is the most common device used for measuring plumbing (Figure 3-39). Stick rules are less common for plumbing, but they may serve the purpose (Figure 3-40). Given the choice, pick a quality tape measure with a minimum length of 20 feet.

Levels

Professionals typically own at least four different levels. They are a torpedo level (Figure 3-41), a grade level (Figure 3-42), and standard levels (Figure 3-43) in lengths of 2 and 4 feet. The torpedo sees constant use, while the grade level is used primarily in new installations. The grade level has an adjustable device on it to allow the plumber to maintain a constant pitch on his pipes. With the adjustment set at the desired grade, for example, $1/4$ inch to 1 foot, the plumber can maintain his grade throughout the job. When the level reads level, the pipe is pitched to the selected grade. The 2-foot level is used frequently when setting fixtures. The 4-foot version is used when installing bathtubs and showers.

Screwdrivers

For most home plumbing repairs, you will only need one screwdriver. This screwdriver is actually four screwdrivers in one. It will have multiple heads and sizes for common and Phillips screws (Figure 3-44). You can collect other screwdrivers as you go along. There may be times when you need larger or smaller screwdrivers.

Wrenches

Wrenches account for much of the space in a plumber's toolbox. You may need pipe wrenches for steel pipe, stubborn clean-out plugs, and a host of other hard-to-turn connections. Plumbers generally carry at least three different sizes of pipe wrenches. They are 10-inch, 18-inch, and 24-inch pipe wrenches (Figure 3-45).

When pipe wrenches are not the right choice, chain wrenches can be the answer (Figure 3-46). These specialty wrenches have strong gripping power and can function in small spaces. The chain wrench's cousin is the strap wrench. These wrenches work on a similar principle, but because they employ the use of a cloth strap, they don't mar finishes. This can be a big plus when working with trim plumbing materials.

No plumber's toolbox would be complete without a basin wrench. Basin wrenches are used to get the nuts off of faucets and for any other task when the average wrench can't reach it. Basin wrenches with telescoping bodies are the best for all your plumbing needs (Figure 3-47).

Adjustable wrenches in assorted sizes are mandatory for your tool kit. These common wrenches (Figure 3-48) are the plumber's workhorse. They fulfill every duty from working on faucets to setting toilets. If you attempt to change a screw-in element in your water heater, you will be glad you have an element wrench. A set of tub wrenches makes replacing the washers in some tub valves much easier.

Miscellaneous Tools

There are some other miscellaneous tools you will find very helpful in some of your plumbing work. A good electrical meter capable of reading volts and ohms will be needed on some jobs. For the homeowner, a meter doesn't have to be a top-of-the-line model. Inexpensive meters can save you hours of frustration and could save your life.

A set of sockets and a ratchet will make your life easier in many circumstances. If you have to drill holes, you may find a need for a nail puller. Trying

to cut nails with your drill bit is no fun. A set of wire strippers tends to prevent fingers from being cut. Skinning electrical wire with a knife will work, but one slip and you can get a serious cut. A striker is nice for lighting your torch. While you can substitute for a striker with a match or a cigarette lighter, the striker is safer. Safety glasses are an essential part of your toolbox. These inexpensive items can save your eyesight.

The list of potential tools is too large to go through individually. With plumbing, anything is possible. For the most part, the tools described above will meet your everyday needs. As you work into being a weekend plumber, you will know if you need a different tool. Now, let's take a look at the types of materials with which you will use the tools.

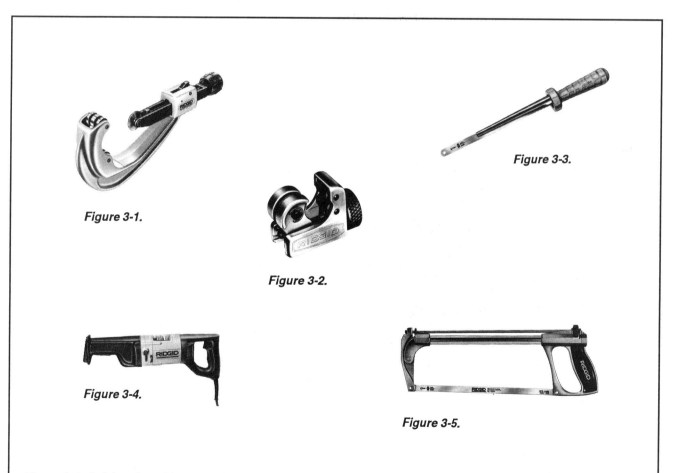

Figure 3-1.

Figure 3-2.

Figure 3-3.

Figure 3-4.

Figure 3-5.

Figure 3-1. Quick-acting tubing cutters. Courtesy of Ridge Tool Company. Figure 3-2. Midget tubing cutters. Courtesy of Ridge Tool Company. Figure 3-3. Jab saw. Courtesy of Ridge Tool Company. Figure 3-4. Reciprocating saw. Courtesy of Ridge Tool Company. Figure 3-5. Non-adjustable hacksaw. Courtesy of Ridge Tool Company.

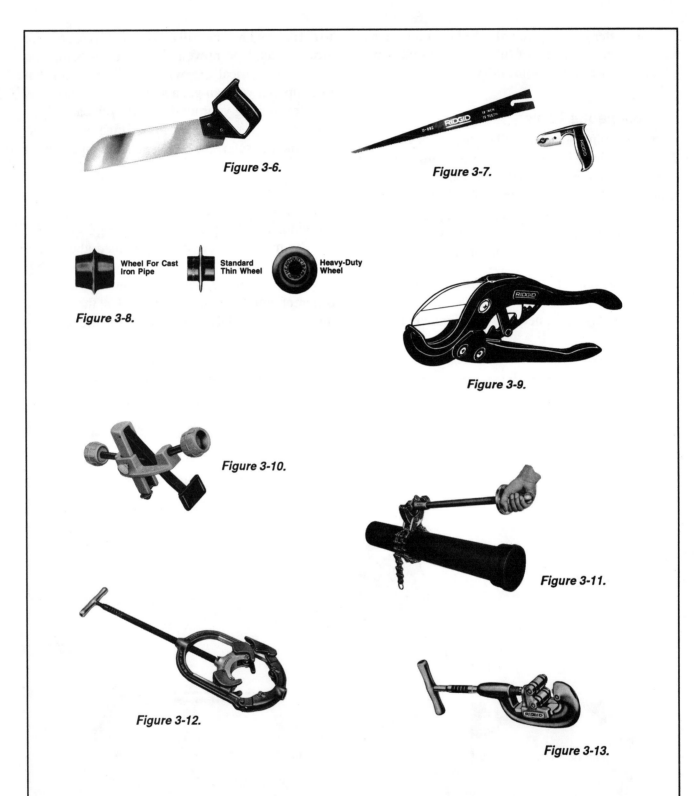

*Figure 3-6. Universal saw. Courtesy of Ridge Tool Company. **Figure 3-7.** Compass saw blade and handle. Courtesy of Ridge Tool Company. **Figure 3-8.** Pipe cutter wheels. Courtesy of Ridge Tool Company. **Figure 3-9.** Plastic pipe and tube cutter. Courtesy of Ridge Tool Company. **Figure 3-10.** Internal tubing cutter. Courtesy of Ridge Tool Company. **Figure 3-11.** Ratchet snap-cutter. Courtesy of Ridge Tool Company. **Figure 3-12.** Hinged pipe cutter. Courtesy of Ridge Tool Company. **Figure 3-13.** Wide roll pipe cutter. Courtesy of Ridge Tool Company.*

Figure 3-14.

Figure 3-15.

Figure 3-16.

Figure 3-17.

Figure 3-18.

Figure 3-19.

Figure 3-20.

Figure 3-21.

Figure 3-14. Two-handle heavy-duty pipe cutters. Courtesy of Ridge Tool Company. *Figure 3-15.* Closet auger. Courtesy of Ridge Tool Company. *Figure 3-16.* Flat-tape snake. Courtesy of Ridge Tool Company. *Figure 3-17.* Hand-operated spring snake. Courtesy of Ridge Tool Company. *Figure 3-18.* Hand-held drain gun. Courtesy of Ridge Tool Company. *Figure 3-19.* Brace-mounted drain-cleaning machine. Courtesy of Ridge Tool Company. *Figure 3-20.* Foot-operated drain-cleaning machine. Courtesy of Ridge Tool Company. *Figure 3-21.* Standard drain-cleaning machine. Courtesy of Ridge Tool Company.

Figure 3-23.

Figure 3-24.

Figure 3-25.

Figure 3-26.

Figure 3-27.

Figure 3-28.

Figure 3-29.

Figure 3-23. Right angle drill. Courtesy of Ridge Tool Company. Figure 3-24. Pistol-grip drill. Courtesy of Ridge Tool Company. Figure 3-25. Hole saw kit. Courtesy of Ridge Tool Company. Figure 3-26. Acetylene torch kit. Courtesy of Ridge Tool Company. Figure 3-27. Propane torch kit. Courtesy of Ridge Tool Company. Figure 3-28. Stepladder. Courtesy of Ridge Tool Company. Figure 3-29. Extension ladder. Courtesy of Ridge Tool Company.

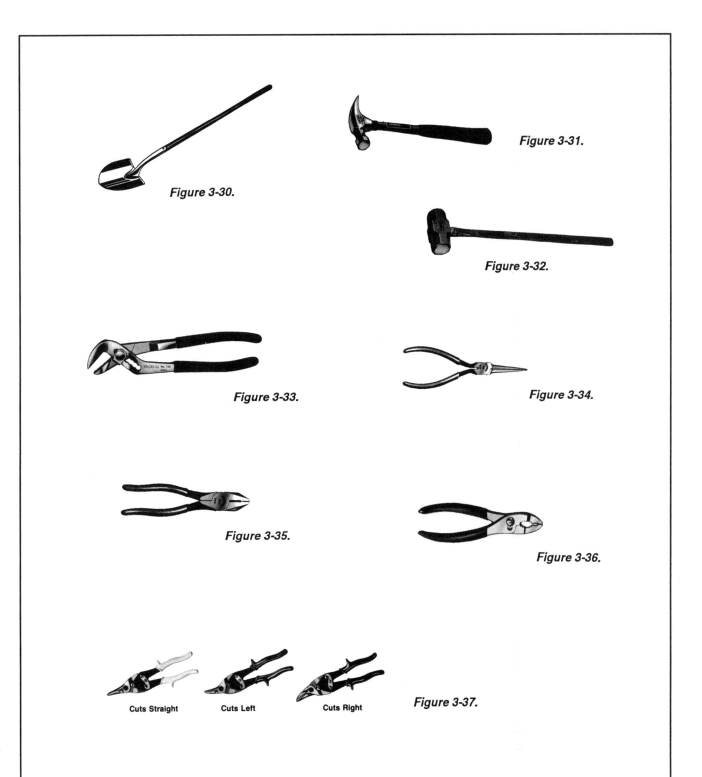

Figure 3-30. Round-point shovel. Courtesy of Ridge Tool Company. Figure 3-31. Claw hammer. Courtesy of Ridge Tool Company. Figure 3-32. Sledgehammer. Courtesy of Ridge Tool Company. Figure 3-33. Angled pliers. Courtesy of Ridge Tool Company. Figure 3-34. Needle-nose pliers. Courtesy of Ridge Tool Company. Figure 3-35. Lineman's pliers. Courtesy of Ridge Tool Company. Figure 3-36. Slip-joint pliers. Courtesy of Ridge Tool Company. Figure 3-37. Aviation snips. Courtesy of Ridge Tool Company.

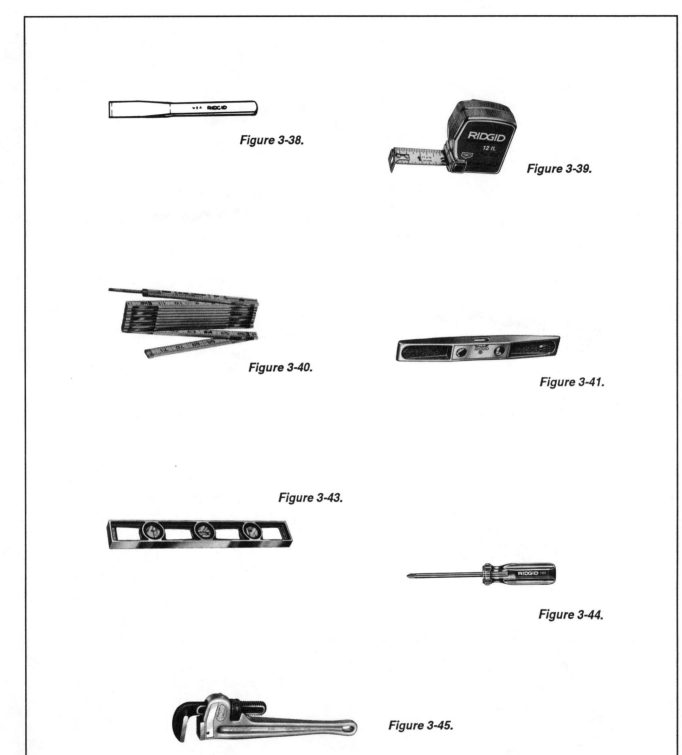

Figure 3-38.

Figure 3-39.

Figure 3-40.

Figure 3-41.

Figure 3-43.

Figure 3-44.

Figure 3-45.

Figure 3-38. Cold chisel. Courtesy of Ridge Tool Company. **Figure 3-39.** Tape measure. Courtesy of Ridge Tool Company. **Figure 3-40.** Extension ruler. Courtesy of Ridge Tool Company. **Figure 3-41.** Torpedo level. Courtesy of Ridge Tool Company. **Figure 3-43.** Standard level. Courtesy of Ridge Tool Company. **Figure 3-44.** Phillips screwdriver. Courtesy of Ridge Tool Company. **Figure 3-45.** Pipe wrench. Courtesy of Ridge Tool Company.

Figure 3-22.

Figure 3-42.

Figure 3-46.

Figure 3-22. Sewer bag. **Figure 3-42.** Grade level. **Figure 3-46.** Chain wrench.

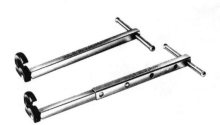

Figure 3-47. Basin wrench. Courtesy of Ridge Tool Company.

Figure 3-48. Adjustable wrench. Courtesy of Ridge Tool Company.

4
Plumbing Fixtures and Materials

When you begin to work with plumbing, it is essential to use the proper materials. There are code restrictions on the use of many materials, and compatibility problems exist for some products. In order to be successful, you must know what materials to use for various jobs. This chapter explains the materials and fixtures you may need to use.

To avoid making this chapter read like a cryptic code book, the information is presented in easily understood terms. When appropriate, I will expand on the reasons for specific code requirements. The first part of this chapter describes basic residential code requirements.

MATERIALS FOR WATER SUPPLY PIPING

The type of pipe used for the distribution of potable (drinking) water must be resistant to corrosive action. If the ground contains solvents, fuels, organic compounds, or other detrimental materials, you may have a serious problem. The ground may have to be tested through a chemical analysis to determine a suitable material for your water service pipe.

Water distribution pipes should not contain more than a maximum of 8% lead. When soldering copper water tube or pipe, you must use a solder with low lead content. There are several types of solder available with a lead content of 5% or less. You should not use 50/50 solder for potable water pipe connections. This solder is still sold and may be used on heating systems, but it should not be used on potable water supply pipes. When buying your solder, request a 95/5 or lead-free solder.

Technically, most residential copper water supply pipe should be called *copper tubing*. For simplicity's sake, all copper tube and pipe is referred to here as *pipe*. All water service pipe installed outside of your home and underground should be rated for a minimum working pressure of 160 pounds per square inch (psi) at 73.4° F. If the water pressure in your area exceeds 160 psi, a pipe rated for the higher pressure is required. When the water service pipe entering your home is plastic, it should terminate within 5 feet of the point of entry.

The most common approved materials for water service pipe are as follows:
- Copper, normally type "L" or "K"
- Chlorinated polyvinyl chloride (CPVC) plastic pipe
- Polybutylene (PB) plastic pipe
- Polyethylene (PE) plastic pipe
- Polyvinyl chloride (PVC) plastic pipe

Brass and galvanized steel pipe are still approved materials, but they are outdated. Their tendency to develop leaks leaves them a material of the past. Copper is still the most common material in many areas, but PB and PE plastic pipe are taking a major share of the market for water service pipe. The plastic pipe is less expensive, is easier to work with, and gives years of uninterrupted service.

CPVC and PVC are not as well accepted as the other plastic pipes for water distribution. These plastic pipes tend to be brittle. Most professional plumbers use copper, PB, or PE pipe for water service installations and repairs.

Water Distribution Pipe

The pipe delivering your potable water within the house may be copper, PB, brass, CPVC, or galvanized steel. Brass pipe is a thing of the past, and galvanized pipe is a very poor choice. It will rust and cause many problems. The threaded joints will rust through, causing leaks. The interior will collect and encourage the growth of compounds that will ultimately close up the pipe. This retards water pressure and requires the replacement of the pipe.

Polybutylene (PB) is gaining popularity fast. It is inexpensive, resistant to splitting in freezing conditions, and easy to install. Copper is the old workhorse. Copper pipe is favored by old-school plumbers and many homeowners. It resists build-ups, is relatively easy to work with, and is a proven performer. Pipe used for potable hot water should be rated for 80 psi at 180° F.

Drain and Vent Materials

Your drain, waste, and vent (dwv) system may consist of many types of materials. The following list includes the various types of pipe you may incorporate into your dwv system:

- Acrylonitrile butadiene styrene (ABS) plastic pipe
- Aluminum tubing (not common in residential applications)
- Brass pipe (outdated)
- Cast iron pipe
- Copper (Types: K, L, M, or DWV) pipe
- Galvanized steel pipe (not a good idea)
- Lead pipe (an unlikely choice)
- Polyvinyl chloride (PVC) plastic pipe (dwv type)

Many older homes contain any of these pipes, except aluminum tubing. Cast iron, galvanized steel, and copper were all common materials in houses of yesteryear. Lead pipe can be found in some older homes, especially in traps and closet bends (the fitting on which your toilet sits). Today, most new homes have PVC or ABS drains and vents. My personal preference is ABS, but PVC is the most prevalent. It is less expensive and more common than ABS.

ABS pipe offers many advantages over PVC. It is easier to cut with a handsaw. It is less prone to leaks, and it is more flexible. Cold weather does not make ABS as brittle as PVC. I have seen trucks run over ABS and not break it. In contrast, I have dropped PVC on a cement floor and had it crack. With PVC, it is important to remove all dirt, mud, and other elements before making your solvent joint. Without the proper cleaning, priming, and gluing, PVC is known for its leaks. ABS is much more user-friendly.

Underground DWV System Materials

When installing your system below ground, you may use the following materials:

- ABS
- PVC
- Cast iron
- Copper (type K or L)

These materials are approved for use with the internal plumbing system. When it exits the house, your building drain becomes a sewer. There are additional materials that may be used in sewer piping, but the above materials should be all you need to know about. Unless you are a professional plumber, there are few times you would wish to use anything else. Of these materials, copper is the one used the least in modern plumbing systems. Cast iron pipe is still used, but it has been practically replaced with ABS and PVC.

Fittings, Valves, and Nipples

Fittings should not be made in such a way as to obstruct the flow in the piping. If threaded fittings

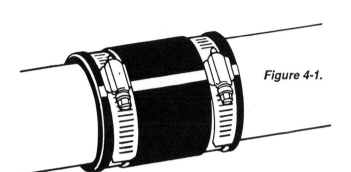

Figure 4-1.

Figure 4-2.

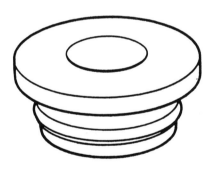

Figure 4-3.

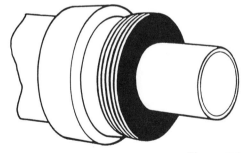

Figure 4-4.

Figure 4-1. Rubber coupling. Courtesy of Fernco. **Figure 4-2.** Reducing-style rubber coupling. Courtesy of Fernco. **Figure 4-3.** Doughnut adapter. Courtesy of Fernco. **Figure 4-4.** Doughnut adapter on hub. Courtesy of Fernco.

are used, they must be of the recessed-drainage type. The fittings should be of the same material as the pipe and designed for the use as installed. Valves must be of an approved type and compatible with the other materials being used. This same rule applies to manufactured pipe nipples.

Some Special Material Requirements

Copper or tubular brass traps and tail piece fittings should be at least a 17-gauge material. Floor flanges for toilets must meet certain requirements. For thickness, brass should be at least 1/8 inch thick; plastic flanges should be at least 1/4 inch thick. If you are working with a cast iron flange, there should be a minimum of a 2-inch caulking depth. When installing these flanges, you should use corrosion-resistant screws or bolts.

Clean-out plugs in your drains should be either brass or plastic. The plugs should have raised square heads or countersunk square heads where the protrusion of a raised head might cause a safety hazard. Unless there is a hazard created by the raised head, raised-head clean-outs are the preferred choice.

FIXTURES

All plumbing fixtures must be of an approved type. They are required to have smooth, impervious surfaces and be free of defects. Fixtures may not have concealed fouling surfaces. The most common approved types of fixtures include:

- Stainless steel
- Enameled cast iron
- Porcelain enameled steel
- Fiberglass
- Vitreous china
- Plastic

Basic homes have similar plumbing fixtures. They include water closets (toilets), sinks, lavatories, water heaters, bathing units, and assorted other fixtures. Many of these fixtures will interchange with existing plumbing. You may be required to make minor modifications, but the basic fixtures, in the proper size, should replace most existing units.

COMPATIBILITY

Since it is very likely you will encounter several different types of materials, it is important to understand compatibility. When you begin repairing old plumbing, you will most likely do so with modern materials. With the proper fittings, almost all plumbing materials can be made to join with others. It is when you are not aware of the proper devices that the risk of a problem arises.

Obviously, when you try to join CPVC pipe to copper pipe, you cannot solder the two together. Attempting to connect cast iron drains with plastic pipe requires a special coupling. Knowing what these devices are and how they work will save you much time and frustration. The remainder of this chapter is dedicated to real-world plumbing material applications.

When you are faced with mating new plumbing materials to existing pipe, you will need adapters and couplings. In the following paragraphs, many of these devices are described. Having the proper adapter can turn a tough job into an easy one.

DWV Adapters and Couplings

For the most part, plastic pipe is the standard in today's dwv system. When you need to repair or connect to older pipe, you will need some special fittings. The best universal connector is a rubber coupling. These fittings can be used to connect plastic pipe to cast iron, copper, galvanized steel, other types of plastic pipe, brass, and just about any other material you could name.

These handy couplings slide over the pipes and are secured with stainless steel clamps (Figure 4-1). They are thick and flexible, and make a solid connection. Since they can be used in almost any application, they are making other adapters obsolete in the professional ranks. There are less expensive alternatives, but for the time saved and the elimination of a need to carry extensive inventory, they are very popular. These rubber couplings are available in all the major pipe sizes and can even be purchased in reducing-coupling styles (Figure 4-2). This allows you to join pipes of different diameters with the same coupling.

Male and female adapters are another common choice for most pipes. They can be installed by normal means on each type of pipe and then screwed together to make a compatible connection between the two different types of pipe. These adapters are available in plastic, brass, copper, and other materials.

In addition to these two types of connectors, there are special connectors for cast iron pipe. You may use the described rubber couplings to connect cast iron with other pipes, or you can use any of the following. There are special rings made to insert into the hub of cast iron (Figure 4-3). These doughnut-shaped rings go into the cast iron, and then the other pipe is inserted into the rubber ring (Figure 4-4). This makes a compatible connection. There are also bands made to join cast iron pipe that doesn't have a hub to other pipe. These devices utilize a rubber ring and a stainless steel ring, secured by stainless steel clamps (Figure 4-5).

Water Distribution Adapters

Male and female adapters are the most common fittings used to join different types of water pipes. These may be plastic, nylon, copper, or some other approved material.

ILLEGAL USES

While a plumbing material may be approved for one use, it may be illegal in a different situation. For example, you may use galvanized steel pipe above ground but not below ground. When you enter the technical chapters, I will point out any illegal uses of various plumbing materials.

THE WRAP-UP

This chapter covers the basics of approved residential materials. If you require a more detailed description, contact your local code enforcement office. The code enforcement office will be happy to help you obtain a current local code book. Now, it is time to get technical. Let's move on to Chapter 5 and explore the plumbing in your kitchen.

Figure 4-5. Hubless band.

5
Kitchen and Appliance Plumbing

As it relates to plumbing, your kitchen is the second most important room in your home. Only bathrooms overshadow kitchens in the field of plumbing dominance. If your kitchen plumbing is ailing, this is the chapter to show you how to cure it.

Take a moment and think of all the plumbing-related fixtures, appliances, and piping in your kitchen. The most obvious plumbing consideration is the kitchen sink and faucet. Do you have a garbage disposer or a dishwasher? Is your refrigerator equipped with an icemaker? Does your sink have cut-off valves under it for the water supply pipes? Is your sink's trap vented? When you start looking at these questions, you may realize you don't know your kitchen very well.

When you use a room daily, it is easy to take it for granted. If you were washing dishes and your faucet started spraying water all over the room, would you know how to cut the water off quickly? Garbage disposers jam periodically, do you know how to free your disposer's impeller? These are only a few of the many procedures you will learn in this kitchen chapter.

This chapter is going to teach you to repair and replace your kitchen's plumbing. The chapter is broken down into specific categories for each of the plumbing jobs discussed. First, you will learn to repair the object. Then replacement techniques are described. All you have to do is look for the sub-

heading of the plumbing item you wish to work with. From that point on, the chapter will lead you step by step to the end of your plumbing endeavor.

THE KITCHEN SINK

Kitchen sinks incorporate the use of many supporting plumbing articles and miscellaneous accessories. Each potential problem has its own subheading and explanation of cause and cure. In addition, I have rated the degree of difficulty for many of these tasks. You are about to take off your reading glasses and pick up your toolbox.

What's Under My Kitchen Sink?
There are many parts contributing to your sink's ability to function. You will find most of these parts under the sink. They range from simple washers to confusing waste assemblies. Before we look at the ways to repair and replace items related to the kitchen sink, let's identify the parts you may encounter.

Basket Strainers
Basket strainers are the pieces filling the hole in the bottom of your sink when you don't have a garbage disposer. They are the starting point of the sink's drainage system. When looking from below the sink, basket strainers protrude through the sink and are attached with a large nut or a retainer device.

The strainer tapers in as it points downward and has threads on the end. The threads are designed to accept a 1½-inch slip-nut. There are two flat bars running across each other in the opening of the basket strainer. These bars prevent large items from entering the drainage system.

Sink Tailpieces

Sink tailpieces are frequently chrome-plated metal, but they may be made of plastic. In standard applications, they are 4 inches long and 1½ inches in diameter for a kitchen sink. Sink tailpieces are available in various lengths to make your connections easier. A sink tailpiece is flanged on one end and straight on the other. The flanged end attaches to the bottom of the basket strainer with a tailpiece washer and a slip-nut.

Slip-Nuts

Slip-nuts are the nuts used to connect many of your waste pipes before they reach the trap arm of the drainage system. Slip-nuts may be metal or plastic and are usually 1½ inches in diameter under a kitchen sink. The nuts are tightened by turning them clockwise and loosened by turning counterclockwise. Normally, there is no need to apply Teflon® tape or pipe compound to the threads before tightening the slip-nuts. Slip-nuts depend on washers to make a watertight seal.

Slip-Nut Washers

Slip-nut washers are used with slip-nuts to make watertight joints. The washers may be rubber or plastic. Some washers are flat and others are beveled. When using a beveled washer, the flat side should rest against the slip-nut, with the bevel pointing into the downward flow of the drainage pipe. As the slip-nut is tightened, the washer is compressed and expands outward. This action forms a watertight seal at the connection.

Continuous Wastes

If your sink has more than one bowl, it may have a continuous waste. Continuous wastes are tubular drains connecting all sink bowls to a common trap.

There are two basic types of continuous-waste designs found in average homes. One is an *end-outlet waste* (Figure 5-1), and the other is a *center-outlet waste* (Figure 5-2). The type used on your sink is determined by the location of the trap. Continuous wastes may be made of plastic or metal.

Tailpiece Extensions

Tailpiece extensions come in various lengths and are equipped with threads and a slip-nut on one end and a straight edge on the other (Figure 5-3). When the tube from the continuous waste is not long enough to reach the trap, these extensions can be added to obtain the desired length.

Traps

The trap under your sink performs multiple duties. The primary function of the trap is to form a water seal between the drainage system and your home's open air. The water seal prevents sewer gas from escaping up the drain of your sink. Without a trap, your home could be filled with unpleasant and potentially dangerous odors and sewer gas. Traps also help prevent foreign objects from entering the home's main drainage system. Depending on where you live and the age of your home, you could have one of several types of traps.

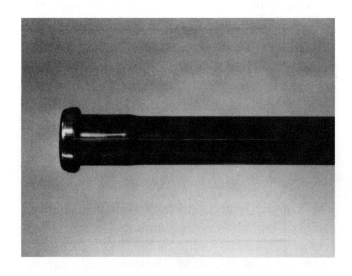

Figure 5-3. Tailpiece extension.

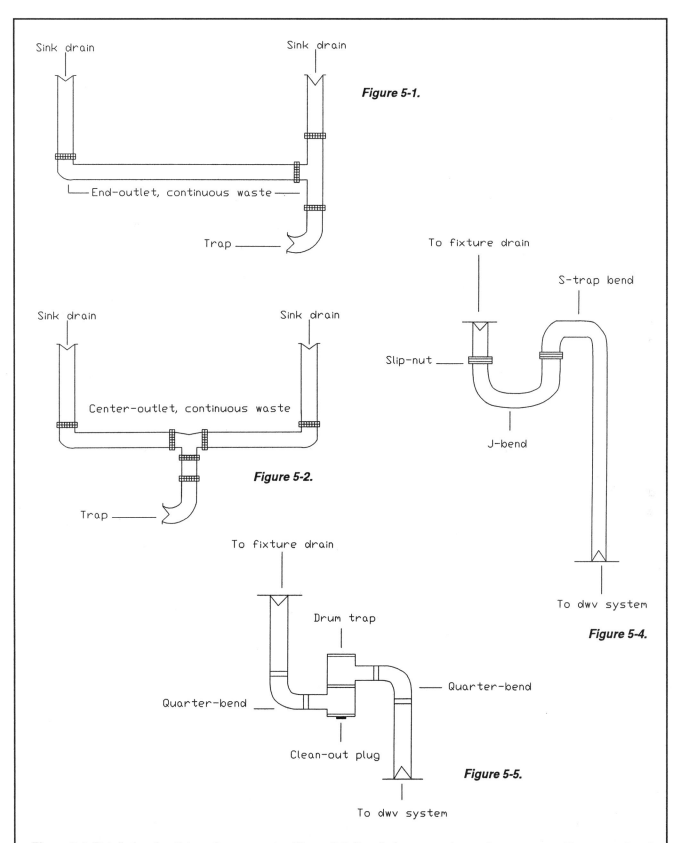

Figure 5-1. Detail of end-outlet continuous waste. **Figure 5-2.** Detail of center-outlet continuous waste. **Figure 5-4.** Detail of "S" trap. **Figure 5-5.** Detail of drum trap.

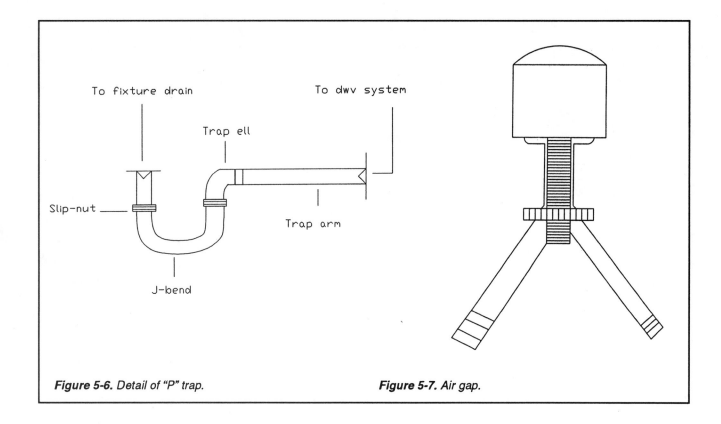

Figure 5-6. Detail of "P" trap.

Figure 5-7. Air gap.

The three most common types of traps are *"S" traps, "P" traps*, and *drum traps*. In most jurisdictions, "S" traps (Figure 5-4) and drum traps (Figure 5-5) are now prohibited by the plumbing code. Drum traps may be approved by the code enforcement office for special applications. A "P" trap (Figure 5-6) is the preferred trap for vented plumbing systems. Your sink's drain connects to the trap with a trap adapter and a slip-nut.

Garbage Disposers and Dishwasher Drains

If your sink is equipped with a disposer, the disposer hangs under the sink and acts as the drain for the bowl to which it is attached. A garbage disposer has a connection on the side to accept the drainage from a dishwasher. If you see a black rubber hose running under your sink, it should be going to your dishwasher. In newer homes, the hose runs from its connection to the drain upward to an air gap (Figure 5-7). It attaches to the larger portion of the wye configuration, and there is a smaller hose running from the wye to the dishwasher.

FAUCETS

Most kitchen faucets are deck-mounted on the sink (Figure 5-8). This means they sit on the flat ledge of the sink with the connections extending below the sink. They are connected to the sink with mounting nuts and may be equipped with a spray attachment.

Supply Tubes

Supply tubes may be brass, chrome-plated brass, or polybutylene. Some supply tubes are made from very flexible, ribbed materials. These tubes connect to the faucet with supply nuts and extend down to the water valves (Figure 5-9). When they connect to the valves, they may do so by a soldered connection or a compression fitting. Compression fittings are the normal method of connection.

Water Valves

Your water valves may be *angle stops* (Figure 5-10), *straight stops, stop-and-waste valves* (Figure 5-11), or almost any other type of shut-off valve. Straight and angle stops are the most common.

These valves are connected to the main water supply pipes under the sink. They may use soldered connections, compression fittings, or screw fittings. Compression and solder connections are the most common. You may have additional valves under the sink. If you have a dishwasher or an icemaker, there should be separate valves for each of them.

MY SINK WON'T HOLD WATER

Kitchen sinks come in all shapes, sizes, and materials. One thing they all have in common is the potential to leak. Some of these leaks escape the plumbing system and saturate the cabinet. More often, the water mysteriously disappears from the sink bowl without a trace. This is the first problem we will troubleshoot and repair.

If you fill your sink with water to do dishes, it should maintain the water level indefinitely. If the water leaves the sink bowl during your dish washing, you probably have a defective basket strainer. Troubleshooting this plumbing problem is simple.

Place paper towels in the base of the sink cabinet and under the kitchen sink. Fill the sink with water. If the leak is very slow, mark the water level with a piece of tape or some other reference point. Give the sink some time to lose its water. When the water level has dropped, check the paper towels. If the water level has receded significantly, the towels will be wet if the water is leaving the plumbing system.

If the towels remain dry, and they usually will, you have a bad basket strainer. In many cases, this repair is as simple as buying a replacement strainer for your sink's bowl. If only the strainer portion is bad, you do not have to remove the strainer housing from the sink, only the basket portion. There are times when you cannot locate a compatible basket to match your existing drain. In these circumstances, you must replace the entire unit. This procedure is explained on page 48.

After testing the sink, if the paper towels are wet, you have a more serious problem. Wet towels indicate water is escaping the drain and running into

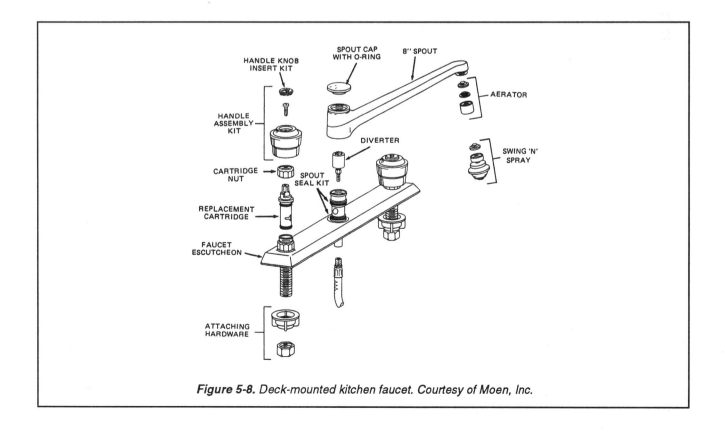

Figure 5-8. Deck-mounted kitchen faucet. Courtesy of Moen, Inc.

the cabinet. This is typically caused by a poor seal around the lip of the basket strainer's housing. When this is the case, you must reseal or replace the existing basket strainer assembly. This is not a difficult job with the proper tools and materials.

To replace the basket strainer assembly, you will need plumber's putty, pipe compound or Teflon® tape, and your tools. The replacement begins under the sink. Basket strainers use two different mechanisms to attach to the sink. Most of them use a large nut, about 2 inches in diameter. Some of them use a flange with three pressure points. These pressure points are threaded rods applying pressure between the flange and the bottom of the sink.

Once you have determined which type of assembly you have, you must loosen the mounting mechanism. For the ones with a nut, you turn the nut counterclockwise. If you don't have a large set of tongue and groove (t&g) pliers, this can be a challenge. In the absence of a suitable wrench or pliers, you can get by with a screwdriver and hammer. The proper wrench is often called a *spud wrench*, but most plumbers use t&g pliers. If you must use a screwdriver and hammer, do it this way. Place the screwdriver bit on one of the nut's lugs. Then tap the screwdriver handle with the hammer. After a while, you will be able to turn the nut with your hand.

The pressure-point type is much easier to remove and install. These units often have flared knobs for you to turn. Some of them require you to turn the threaded portion with a common screwdriver. In either event, this is a much easier style to work with. After the retaining nut or flange is loose, you have it made.

With the hard part done, remove the slip-nut attached to the smaller threaded portion of the basket strainer assembly. Now the drain assembly should lift out of the sink. Remove the old putty around the hole in the sink. This can be done with your fingers or a screwdriver.

Reach into your plumber's putty and pull out a small handful. Roll it in your hands to form a long "snake" of putty. Place the putty around the lip of your new drain assembly (Figure 5-12). Now, push the drain assembly into the hole in the sink; the putty should spread out with the pressure you apply.

If your new strainer is the type with threaded pressure points, it is really much easier to work with. Crawl under the sink and place the fiber washer that came with the drain over the threaded portion until it touches the bottom of the sink. Next, install the retainer nut or flange by reversing the procedure you used in removing the old one. With this accomplished, align the sink tailpiece and attach it to the drain assembly with the slip-nut.

Tighten all the nuts or pressure points until they are firm. Now, fill the sink with water. Follow the same test procedures you used in troubleshooting the problem. Test the sink at least twice. If you are sure nothing is leaking, congratulations. You have successfully completed your first plumbing job.

SPRAY HOSES

Spray-hose attachments are a frequent cause of aggravation in the kitchen. If your faucet incorporates the use of a spray attachment, you can expect to replace it at some time in the future. The list of potential problems includes drips, leaks, breaks in the hose, and sediment deposits in the spray head. None of these defects is particularly expensive or complicated to repair. In most instances, replacing the entire hose assembly is the simplest way to correct deficiencies in the unit (Figure 5-13).

While it may be possible to repair some types of spray heads, it is not cost-effective or feasible. When your spray head does not function properly, you should replace the entire head. If this is not possible, you can replace the hose and spray head. It can be very difficult to find a head for some spray attachments. When a replacement head is not available, change the entire assembly.

If your spray hose is cracked, you should replace the entire unit. Unless your assembly is unusual, it is always best to replace the whole piece when a problem occurs. The cost of the assembly is under $10 and the job is simple. The most difficult aspect of replacing these units is getting to the connection in the base of the faucet.

Figure 5-9.

Figure 5-10.

Figure 5-11.

Figure 5-9. *Supply tube and straight stop valve.* **Figure 5-10.** *Angle stop valve.* **Figure 5-11.** *Stop-and-wlaste valve.*

The best tool for this job is a basin wrench. When you have one of these, the job goes very quickly. The only type of material you will need in addition to the assembly is some type of pipe compound or Teflon® tape. (From this point on in the book I will refer only to pipe compound, but Teflon® tape can be used in its place if you prefer.) Most kitchen faucets with spray connections will accept any universal spray assembly. It is rare that you will need a special brand. When you are ready to start, put on a pair of safety glasses. The job requires you to lie under the sink and look up at the connection. It is not uncommon for rust or other foreign objects to fall into your eyes while working.

With your safety glasses on, crawl under the sink. Look at the base of the faucet where the hose is connected. There is a small brass nut-type arrangement at the end of the hose for your basin wrench to attach to. Put the wrench on the nut-type fitting and turn it counterclockwise. The threaded portion should come out fairly easily. When the hose is free, you simply pull the spray head from above the sink to remove the assembly.

With the old one out, feed the new hose through the hole and into the cabinet. Some units are packaged

without the head attached to the hose. If this is the case, all you have to do is screw the head onto the hose. You should not have to put pipe compound on these threads, but you may find it necessary if the head leaks when tested. Next, apply pipe compound evenly to the threads on the end of the hose where it attaches to the faucet. Now, with your fingers, begin screwing the fitting into the connection on the base of the faucet. Be careful not to cross-thread the fitting. This light-weight brass will damage easily if cross-threaded. When you feel the fitting is snug, finish tightening it with the basin wrench. Don't over-tighten it. You can test for leaks and tighten as needed.

At this point, you are ready to test the installation. Turn the faucet on and depress the handle on the spray head. When possible, have someone run the sprayer while you inspect the connection. If you are working alone, a rubber band can be placed over the handle to force the sprayer to continue producing water. Visually inspect your connection and wipe the fitting with a paper towel. There are times when a leak cannot be detected visually but will show up on the paper towel. If there is a leak, tighten the fitting until it stops.

Figure 5-12. Installing putty on a basket strainer. Model: Andrew E. Wallace.

Figure 5-13. Kitchen spray hose and head. Model: Andrew E. Wallace.

As long as you did not cross-thread the fitting, the leak should stop as you tighten the fitting. If water seems to spray from the fitting, you probably cross-threaded it. If you did, you must remove the fitting and replace the entire hose. Cross-threading will not be a problem if you run the fitting into the connection by hand. This is one of the simplest tasks you can take on in the kitchen.

REPAIRING YOUR FAUCET

Kitchen sink faucets may have one handle or two. They can have washers, cartridges, or even ball assemblies controlling water flow. Depending on the brand and style, repairing your faucet can be easy or extremely complex. Before you begin to work on your faucet, determine what brand of faucet you have. This should be clearly marked on the faucet. There are so many different faucets available, this book cannot detail the repair of each one. I will give you the basics for repairing the most common types of faucets. With these basic principles, you may be able to apply them to your faucet even if it is not adequately described here.

Two-Handle Faucets with Washers

These traditional faucets are relatively simple to work on. The two most common problems are a drip from the spout or water running up from around the handles. To repair these problems, you will need your toolbox and compatible repair parts.

The first order of business is cutting off the water supply to the faucet. This is normally done at the shut-off valves under the sink. If your sink does not have valves under it, you will have to cut off the main water valve to your home's plumbing system. These valves are usually found in basements or closets. When you think the water is cut off, test the faucet to be sure. Open the handles and see if there is any water pressure. Old valves do not always work, and you may still have water at the faucet.

You cannot repair the faucet until there is little to no water pressure at the fixture. Once the water is off, remove the caps on the faucet handles. These caps

cover the screws holding the handles on and may be removed with a knife or screwdriver. When you have access to the screws, turn them counterclockwise. Once the screws are removed, lift the handles off the stems. With old faucets, you may have to use force to remove the handles. Normally, prying with a screwdriver will work. In extreme cases, you may need a handle-pulling tool.

With the handles off, turn the retainer nuts counterclockwise with an adjustable wrench. When they loosen, the stems will start to come out as you turn the wrench. This is when you will find out what type of parts you need. In faucets using washers, you will see the stem with a washer on the end of it. The washer is held in place by a screw. These screws are often deteriorated and will fall into pieces when you attempt to turn them.

Attempt to remove the screw and the washer. If you are able to do this, take the entire stem to your local parts supplier for comparison and acquisition of the new parts. Replacing the washer or the stem should stop water dripping from the spout. If your problem is water escaping around the handle, you will need to replace the "O" rings on the stem. If the part is not repairable, you may have to find a replacement stem for your faucet. To put the faucet back together, reverse the steps you took when dismantling it.

The parts for this type of repair should cost less than $10, unless you must replace both stems. If both sides warrant replacement, your cost might be around $20.

Replacing the Faucet Seats

These faucets have seats in them for the washers to seal against (Figure 5-14). The seats can be replaced using a *seat wrench*. If your faucet seats are bad, you can sometimes make a temporary repair by sanding them with sandpaper. This removes rough spots and voids that allow water to seep past the washers. To replace the old seats, turn the seats counterclockwise with the seat wrench. Install the new seats by reversing the procedure. By the time your faucet seats are ruined, you should consider replacing the faucet.

Two-Handle Faucets with Cartridges

The basic stem/cartridge removal principles are the same for these faucets. Once you have the stem/cartridge out of the faucet, the entire cartridge will need to be replaced to stop a dripping spout. Replacing worn "O" rings will eliminate the water running up around the handle. Replacement cartridges are available for many brands of faucets at less than $10 each.

Single-Handle Faucets

The two most common types of single-handle faucets use either a cartridge or a ball assembly. Without a doubt, the cartridge type is the easiest to repair. These repair parts (Figure 5-15) are sold in kits. For the cartridge type, you can expect to pay between $12 and $22 for the replacement cartridge. The ball-type kits should cost less than $10.

Replacing Single-Handle Cartridges

Cut off the water and remove the cap covering the screw on the faucet handle. By removing the handle, you should see a small metal clip. Be certain the water is off before removing this clip. If there is pressure at the faucet when the clip is removed, the cartridge can become a small missile. With the water off, remove the clip with a pair of pliers. Now, with your t&g pliers, pull up on the cartridge. It should come out without much fuss. Replace the old cartridge with a new one, replace the clip, and put the handle assembly back together again. This job is very simple and requires little time.

Replacing Ball Assemblies

Working with the ball assembly type of faucet can be nerve shattering. There always seem to be springs flying and parts going astray before you know what happens. To remove the handles of most of these types of faucets, you will need a hex wrench. These wrenches are often included in the repair kit.

With the water cut off, loosen the hex screw in the front of the faucet. To remove the handle, tilt and lift it off. You will see a rectangular stem protruding from the dome-shaped retainer. Unscrew the re-

tainer by turning it counterclockwise. When it is removed, you will see the ball assembly and all its many springs, "O" rings, and assorted parts. Your best bet is to replace the entire assembly. Only experienced plumbers can easily tell which individual parts to replace, and the cost of replacing the whole assembly is minimal.

Wall-Mounted Kitchen Faucets

If your faucets are mounted on the wall, be careful (Figure 5-16). The wrong moves with these faucets can break the water pipes in your wall. If this happens, you are facing some major expenses. Basically, the repair procedure is similar to the washer-type faucet. The risk of breaking the pipes is high if you must exert much pressure on the faucet to remove the stem. Definitely don't try this repair when you would have to pay a professional plumber overtime wages.

These faucets are outdated and rare, but they are still manufactured. If everything goes smoothly you can repair them on your own. If you run into complications, call a plumber. Breaking the pipe off in your wall will cause you a lot of grief and expense. When I get to the replacement section, these faucets will be included, but I don't recommend homeowners attempting the replacement.

Leaks at the Base of the Faucet Spout

There are times when kitchen faucets leak where the spout meets the base of the faucet. When you think of how often you swing the spout around, it is understandable that it might leak. There is an "O" ring around the base of the spout. This "O" ring becomes worn and allows water to pass by. This is a very simple repair.

With the handles on the faucet turned off, loosen the knurled ring around the base of the spout. When the ring is loose, you can pull the spout out of the faucet housing. In doing so, you will see an "O" ring at the end of the spout. Find an "O" ring of similar size and replace the worn one. Stick the spout back into the housing and tighten the ring. This repair is done, and you didn't even break a sweat.

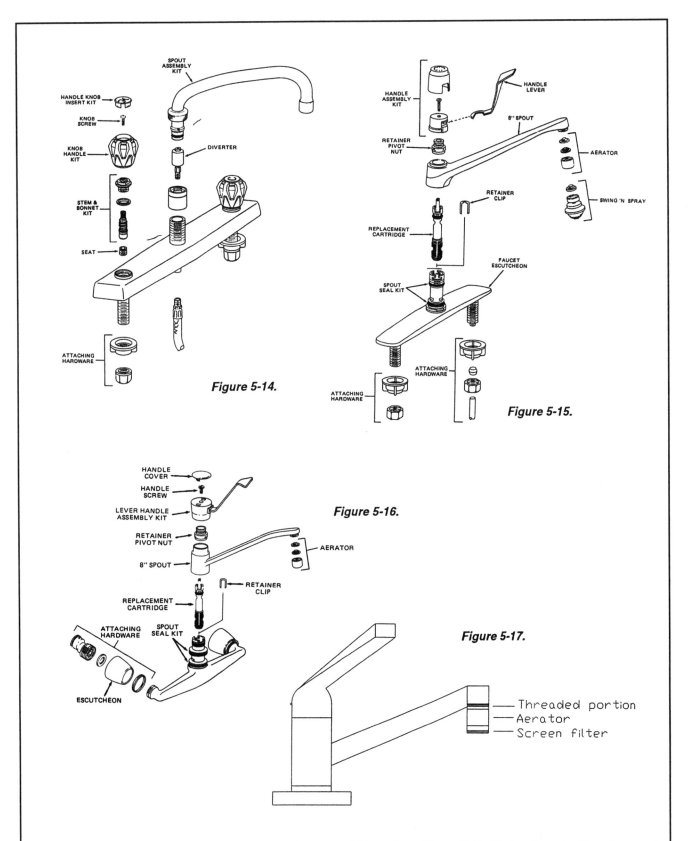

Figure 5-14.

Figure 5-15.

Figure 5-16.

Figure 5-17.

Figure 5-14. Detail of kitchen faucet with seats. Courtesy of Moen, Inc. ***Figure 5-15.*** *Kitchen faucet with replacement cartridges. Courtesy of Moen, Inc.* ***Figure 5-16.*** *Wall-mount faucet. Courtesy of Moen, Inc.* ***Figure 5-17.*** *Detail of an aerator.*

There will be times when it makes more sense to replace your present faucet than to repair it. There will also be occasions when you have little choice in the matter, due to deterioration of critical parts and old faucets. The replacement process is very similar for most types of faucets. Let's take a look at some of them.

REPLACING DECK-MOUNTED FAUCETS

When the faucet you are replacing is deck-mounted, it won't make much difference if it has one or two handles. There are a few brands with different mounting and connection procedures, but most of them are installed in the same manner. We will start with the average faucet replacement.

Getting the Old Faucet Off the Sink

If the job goes in textbook manner, this is a simple process. Unfortunately, old faucets rarely come off in the way you hope they will. When replacing faucets, you should wear safety glasses. There are often rust and other unknown objects capable of hurting your eyes. In faucet replacement, you will be lying under the sink and looking up. This creates a high risk for damage to your eyes.

First, you must be sure the water to the faucet is cut off. Next, you will need your basin wrench. Put the wrench on the supply nuts and turn them counter-clockwise. Some water will drip down the supply tube, but if the valve is holding, the water will stop in a matter of moments. When the supply nuts are loose, they will fall down the supply tube and onto the valve. This stage of the operation usually goes smoothly.

The next step is to remove the mounting nuts holding the faucet to the sink. These are a much thinner material and frequently weaken with rust and age. When you try to turn them with the basin wrench, they may break off. If they do, you have no problem. All you should care about is getting them off the threaded portion of the faucet connection. Sometimes they will not turn. In extreme cases, you will have to cut them off with a hacksaw blade.

Using whatever method is required, remove these nuts.

If the faucet has a spray attachment, the easiest removal method is to cut the hose. This allows the removal of the hose without loosening the fitting. Once the mounting nuts are removed, you simply grasp the faucet and remove it from the sink. This may sound easy, but it can be very frustrating if the mounting nuts do not cooperate. There is not a lot of room to work under the sink, and you can find yourself in some uncomfortable binds.

The New Faucet

Most kitchen sinks have holes drilled in them to accept faucets with an 8-inch center. This means there is 8 inches between the two water inlet connections. Some kitchen sinks use a 6-inch center. This is something you should measure before buying your replacement faucet. There are a few brands of single-handle faucets that will work on either a 6- or an 8-inch center. When you talk with your plumbing supplier, he can advise you on the products available locally.

When you have the proper faucet, replacement is pretty simple. You will need some plumber's putty, but that is about all you need except for the faucet package. If your existing faucet has a spray, be sure the replacement is equipped with one. If the existing faucet doesn't have a spray, avoid models with spray hoses. When you want to eliminate the spray, you can get a cover to hide the extra hole in the sink. When you want to add a spray, you can be looking at a big job.

There are some faucets on the market with the spray hose going through the faucet base. This is the most logical choice for adding a spray where there has not been one. Otherwise, you will have to cut an additional hole for the spray hose to occupy.

Installing the New Faucet

Unpack the faucet and follow the instructions. Generally, the faucet will be packed with a rubber or plastic gasket to cover the base of the faucet. Install this piece on the faucet first. If the faucet doesn't have a gasket, you can make one with your putty. Roll the putty in your hand to make a long line and lay it around the interior base of the faucet.

Now, set the faucet on the sink so its connections penetrate the existing holes. If there is any old putty left on the sink, remove it before installing the new faucet. Press the faucet down until the putty expands beyond the base. If you are using a gasket material, you do not have to exert this kind of pressure. If the faucet has a spray attachment, push the hose through the receptor for it and into the cabinet below the sink. Now it is time to crawl under the sink with your basin wrench and wearing your safety glasses.

Most faucets have a metal ribbed washer that goes onto the threaded connection of the faucet from below the sink. Some have a plastic piece that slides over these connections. Whichever you have, place it on the faucet now. Follow this with the mounting nuts. Run them up the threaded portion until they are hand-tight. Crawl out of the cabinet and see if the faucet is setting straight on the sink. When the faucet is straight, get under the sink and tighten the mounting nuts with your basin wrench.

Many times the existing supply tubes can be connected to the new faucet. Inspect the head of the supply tube. If it appears to be in good shape, place it into the receiver of the threaded connection. Some plumbers use pipe compound on these connections, but they are designed to work without pipe compound. Slide the supply nut up the tube and screw it onto the connection shaft of the faucet. When the supply nut is hand-tight, tighten it with the basin wrench.

If you have a spray hose to attach, prepare the threads of the hose with pipe dope and screw it into the center hole. Tighten it first by hand and then with the basin wrench. You are now ready to test your new installation.

Before cutting the water on, remove the aerator (Figure 5-17) from the faucet. If you fail to remove this, the screen may become blocked by sediment, pipe compound, or a host of other unexpected particles when the water is turned on. Turn on the

valves under the sink and examine your connections to the base of the faucet. If you don't see any leaks, turn the faucet on. Run water from both the hot and cold sides. Inspect your connections again. Wipe the connections with a paper towel to discover any leaks invisible to the eye. With your newfound skill, you should not have any leaks and your job will be done.

If you do have leaks, slowly tighten the fittings until the leaks stop. If the leaks seem to spray water, you may have cross-threaded a connection. In general, that is all there is to changing your own faucet under normal conditions.

SWAPPING OUT WALL-MOUNTED FAUCETS

Wall-mounted faucets are not common, but they can be very difficult to work with. Experienced plumbers often have problems with these old-style faucets. The two most common problems involve the water pipes feeding the faucet. They either break under stress or fall down into the wall when the faucet is removed.

This job is usually best left to professionals. If you decide to attempt the replacement, proceed with care. To replace these faucets, you loosen the nuts where the faucet enters the wall. When you loosen the connections — if you can — the water pipes may fall down or back into the wall. This can create a circumstance in which the wall must be cut open to access the pipes. Twisting on seized fittings can crimp or break the water pipes in the wall. Having the water pipes disappear into the wall can leave your whole house without water until they are fixed.

If all goes well, when the connections are loose you can remove the old faucet. The new faucet will be installed by reversing the procedure used in removing the old faucet.

REPLACING YOUR KITCHEN SINK

If you want to replace your kitchen sink, you can. Modern sinks are made from stainless steel and cast iron. Most houses have stainless steel sinks. The

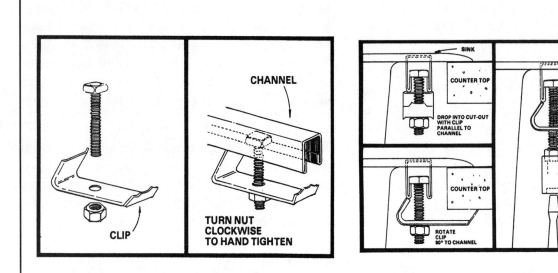

Figure 5-18. *Sink clips. Courtesy of UNR Home Products.*

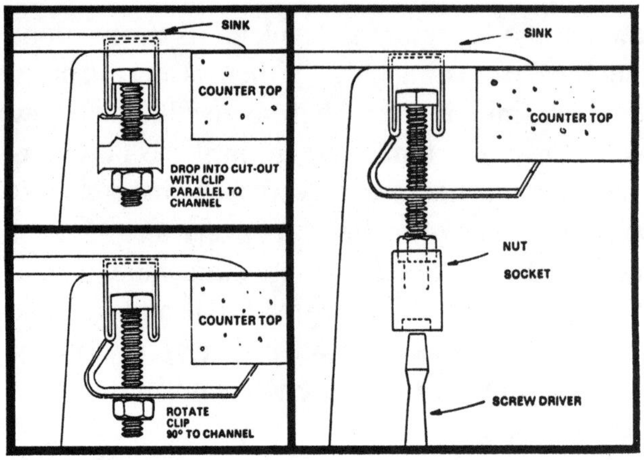

Figure 5-19. *Sink clips in place. Courtesy of UNR Home Products.*

majority of these contain two bowls and are easy to replace. The first criteria is determining what size sink you need. Many people measure the dimensions of their sinks while they are installed in the countertop. This can be risky. Not all sinks have the same size ledges, so you should measure the hole in your counter, not the edges of your sink. This can be done by crawling under the sink with a tape measure and a flashlight.

Once you have the sink you wish to install, you are ready to work. First, you must remove the old sink. Remember to wear safety glasses and follow the instructions throughout this chapter to detach the items affixed to the sink. When the sink is disconnected from the water and waste lines, all you have to do is remove the mounting hardware. These are usually little clips under your countertop and attached to the sink (Figure 5-18).

A screwdriver is normally the only tool required to remove the sink clips. When the clips are gone, the sink should lift out of the hole. If there is a caulked seal between the sink and the counter, carefully cut it with a knife. Remove the old sink and clean the counter. Cast iron sinks don't use clips — their weight is enough to hold them in place. If you have a cast iron sink, you may need help removing it. They are very heavy.

With the old sink out, roll up some plumber's putty. Lay the putty around the opening in the counter and set the new sink into the hole. Press firmly to seat the sink into the putty. Now, crawl under the sink and attach the clips. Most sinks use a clip that slips into a track on the sink and bites into the bottom of the counter as it is attached (Figure 5-19).

When the sink is firmly secure, reinstall the faucet, supply tubes, and drain connections. Other than the uncomfortable working space, this is not a difficult job. If you are using a cast iron replacement sink, follow the same procedures except for the sink clips. The key to success with this task is getting a replacement sink of the proper size.

REPLACING THE AIR GAP

It doesn't get much easier than this. To replace the air gap, you will only need a screwdriver and your t&g pliers. Remove the chrome cover on the air gap with your fingers. Turn the retainer nut counterclockwise until it comes off. Allow the air gap to drop into the base cabinet. Loosen the stainless steel clamps around the hoses with your screwdriver. Pull the existing air gap out of the hoses. Stick the new air gap into the hoses and tighten the clamps. Push the air gap back through the hole in the sink or counter and install the retainer nut. Install the new chrome cover, and you are done.

WORKING WITH THE CONTINUOUS WASTE

The drainage piping between the kitchen sink and the trap is a common site for leaks. Sometimes the slip-nut washers wear out, and sometimes the tubing rusts through. If you have leaks in your continuous waste, you don't need to call a professional plumber. I can show you how to take control of these difficulties and correct them.

Continuous-waste systems are typically connected with slip-nuts. These connections are easy for the average homeowner to work with, and they do not require high-tech tools. For tools, all you will need is a pair of t&g pliers and a large tubing cutter or hacksaw. When your continuous waste is metal, tubing cutters make the job much easier, but a hacksaw will work. If your continuous waste is plastic, a hacksaw is an excellent tool for the job.

If you have a leak around a slip-nut, you probably only need a new washer. These washers are very inexpensive and are easy to replace. Loosen the slip-nut by turning it counterclockwise. When the nut is loose, pull the tubular pipe out of the fitting into which it is inserted. Remove the old washer and install the new one. If it is a beveled washer, place the flat edge into the slip-nut and the beveled edge toward the flow of drainage.

Push the slip-nut to the threads and tighten it with t&g pliers. Under normal conditions, you will not need to apply any pipe dope to the threads. For leaks around the slip-nuts, this procedure usually cures the problem. If the tubular waste has a hole in it, you should replace it. You could put a patch on it, but for the money required, you are better off to replace the tubing.

For this job, you will need the same tools and will perform about the same work. Purchase a new section of tubing and cut it with your tubing cutters or hacksaw to the proper length. You can determine the right length by measuring the old section of tubing. The removal and installation process is the same as in the example above. These are very easy repair and replacement jobs for most people to tackle.

SINK TAILPIECES AND TAILPIECE EXTENSIONS

Replacing sink tailpieces and extensions is very much like working with the continuous waste. The same tools and basic techniques are required. It a simple matter of replacing washers and occasionally tubular tailpiece sections. All you have to do is loosen the slip-nuts and remove the existing tailpiece or extension. Then reverse the procedure to install the new washers or tubular sections.

TRAPS

Traps can be troublesome, but they are not too difficult to work with. If you are replacing a "P" trap with a "P" trap, the job should be fairly easy. It is when you try to alter the style of the trap that the trouble begins. In repair and replacement work, you will almost always be working with the same style trap. It can require major work to switch to a different style. In this section, we will deal with the repair and replacement of similar traps.

Traps may be made of different materials than the continuous waste that connects to them. For this reason, you may need a trap adapter. This adapter allows you to connect a continuous waste to many types of traps made from various materials. It also allows you to connect a metal waste to a plastic trap

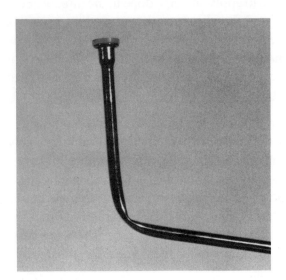

Figure 5-20.

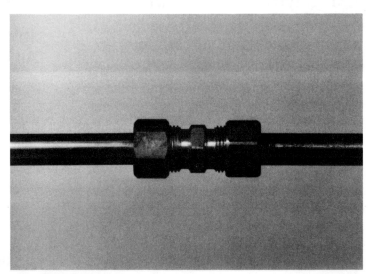

Figure 5-22.

Figure 5-24.

Figure 5-20. *Crimped closet supply tube.* ***Figure 5-22.*** *Compression coupling.* ***Figure 5-24.*** *Dishwasher drainage adapter.*

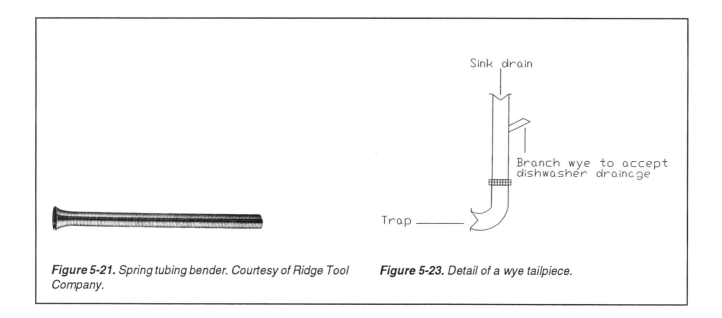

Figure 5-21. *Spring tubing bender. Courtesy of Ridge Tool Company.*

Figure 5-23. *Detail of a wye tailpiece.*

or a plastic waste to a metal trap. The connection between the trap and waste is made with a slip-nut and washer.

Removing the Old Trap

When you replace your trap, you may need to cut the trap arm. The trap arm is the pipe coming out of the wall or floor to the trap. The trap ell may be screwed onto a piece of threaded steel pipe coming out of the wall. You could be dealing with a copper trap arm or a schedule 40 plastic arm. While this may seem perplexing at first, there is no reason for concern.

A majority of traps have three slip-nuts between the trap arm and the sink waste. Loosen the slip-nut where the trap meets the waste. Then loosen the slip-nut in the middle of the trap. When this nut is loosened, water will spill from the trap. Traps are designed to hold water at all times, and when you disconnect the trap you will have to deal with the water. The removal of the "J" bend will disconnect the trap arm from the sink waste. Once the center of the trap is removed, you are ready to determine how you will connect the new trap to the trap arm.

Trap arms coming out of walls usually have "P" traps on them. Trap arms rising from the floor have "S" traps or drum traps. Let's start with "P" traps since they are the most common. The trap ell should

still be connected to the trap arm. If the trap ell appears to be screwed onto the trap arm, try turning it counterclockwise with t&g pliers or a pipe wrench. If the pipe coming out of the wall turns with the trap ell, place pliers or a wrench on it to hold it still. You normally do not want to remove the pipe coming out of the wall. When the trap ell is removed, you should be looking at a $1^1/2$-inch threaded end. The pipe coming out of the wall may be copper or schedule 40 plastic. You can cut these pipes with large tubing cutters or a hacksaw to remove the trap ell.

Installing a New "P" Trap

When the old trap and trap ell are removed, it is time to prepare for the new installation. The new trap will connect to the sink waste with slip-nuts. Some traps come equipped with the slip-nut connection as an integral part of the trap. If you are using a plastic trap, you may need a trap adapter. These adapters fit between the trap and the sink waste.

The next step is to determine how the trap connects to the trap arm. If you are looking at a threaded piece of pipe for the trap arm, you have three options. You can use a threaded trap ell, a female adapter, or a rubber coupling. The threaded trap ell will screw onto the arm, but you should be sure to apply pipe compound to the threads before installing it. The female adapter will screw onto the trap arm and accept the same pipe material from which the trap is made. When using a female adapter, you will also need to apply pipe compound to the threads on the arm.

For example, assume you have a galvanized steel trap arm with threads and a PVC plastic trap. You could screw a PVC female adapter onto the steel threads and glue a piece of PVC plastic into the other end to mate with the trap. This female adapter allows you to make a satisfactory connection between two different types of materials. If you are looking at a fitting with threads inside it, you would use a male adapter. The same principles apply, and you would apply pipe compound to the threads of the male adapter.

In the case of copper pipe, you can solder on screw adapters or use rubber couplings. These rubber couplings can be used on any type of drain pipe. They are effective on steel, copper, bronze, cast iron, and plastic. The only concern here is that there is enough room for the coupling to fit between the wall and the trap fitting. When you have enough space for these couplings, they are the easiest method to use for connecting the old to the new.

If your trap arm is schedule 40 plastic and your trap is of the same material, you can use a regular plastic coupling. These couplings attach to the pipe with a solvent solution. In the case of PVC (white plastic pipe), you should use a cleaner, a primer, and a solvent to make the connection. Many plumbers omit the cleaner and primer and only use the glue, or solvent. This is against the plumbing code, but it is done all the time. To ensure the best joint possible, use the cleaner and the primer.

Black plastic pipe (ABS) is much less finicky than PVC. You can get a very good joint with only the use of a proper solvent. While glue-together connections are not difficult, a rubber coupling is a good choice for the homeowner. Glue joints set up fast. If you make a mistake, you will have to cut the pipe out. With rubber couplings, you can work with them until you get the connection you want.

Installing New "S" Traps

Putting in a new "S" trap is very similar to the "P" trap installation. The major difference is the location of the trap arm. Trap arms for "S" traps come up through the floor. There is typically a slip-nut at the floor level for the "S" trap to fit into. These traps are now illegal by most codes, but they are still available at suppliers to replace existing "S" traps.

For these traps, it is normally only a matter of removing slip-nuts and installing the new trap assembly. You may need tubing cutters or a hacksaw to make the trap tubes the proper length.

Installing Drum Traps

Drum traps and anti-siphon traps are normally illegal except by special permission from the code enforcement office. The trap arm for these traps can come out of the wall or the floor. The replacement procedures for both are essentially the same as for other traps.

REPLACING SUPPLY TUBES

Supply tubes are the small tubes running from the cut-off valves under the sink to the bottom of the faucet. They supply the water to the faucet and are usually a ³/₈-inch tube. They may be made of copper, bronze, plastic, ribbed materials, or a number of other configurations. Replacing these tubes is not difficult if you have a basin wrench.

Most homes are fitted with chrome-plated metal supply tubes. These are the most difficult type for homeowners to replace. When you try to bend these tubes with your hands, they tend to crimp up (Figure 5-20). You can purchase a spring bender (Figure 5-21) to help avoid this problem. Plumbers work with these often enough to bend them without the aid of bender tools. For homeowners, polybutylene or ribbed supply tubes are the easiest to work with.

Removing Existing Supply Tubes

To remove supply tubes you will want a basin wrench and an adjustable wrench. Cut off the water at a point before it reaches the supply tube. This can normally be done at a valve under the sink. Place the basin wrench on the supply nut where it connects to the faucet. Turn the nut counterclockwise. The supply nut will drop down to the valve while remaining on the supply tube. The nut cannot be removed fully unless the supply tube is removed. Then use the adjustable wrench to loosen the nut where the tube enters the valve.

In rare cases, the supply tube will be soldered into the valve. In this case, use your tubing cutters to cut the supply tube about 4 inches above the connection with the valve. With the old supply tube out, you are ready to install the new one.

Installing Supply Tubes

Assuming you have the type of valve where the tube was connected with a nut, you will have to cut the old tube to retrieve the existing nut. You will also need a ³/₈-inch ferrule. For metal supply tubes, the ferrule must be brass. For plastic supply tubes, the ferrule is nylon. Hold the new supply tube head up to the faucet connection. With the bottom of the tube near the valve, mark a place to cut it.

The supply tube should slip into the valve to the bottom of the threads for the nut. If the tube needs to be bent to make the connection, bend it before cutting it. You can cut the supply tube with tubing cutters or a hacksaw. Tubing cutters are much easier to use than a hacksaw for metal supply tubes. When the tube is the proper length, slide the large supply nut on it with the threads toward the faucet connection. Next, slide the smaller nut onto the tube with the threads facing the valve. Now, slide the ferrule onto the supply tube.

Place the supply tube into the valve and tighten the small nut until it is hand-tight. Position the head of the supply tube into the faucet connection and tighten the supply nut until it is hand-tight. If everything looks right, tighten the supply nut with your basin wrench. When it is tight, tighten the smaller nut at the valve with an adjustable wrench. With both nuts tight, turn on the water and test for leaks. Use a paper towel to expose any leaks you cannot see.

Connecting Supply Tubes to Themselves

If you had to cut the supply tube off above the valve, you will need to use a different technique. Purchase ³/₈-inch compression couplings (Figure 5-22). After making your connection at the faucet, connect the tubes with the compression couplings.

Slide the coupling nuts and ferrules over each tube and place the coupling between them. Run the nuts up hand-tight on the coupling. Hold the coupling with an adjustable wrench while you tighten each nut. There is no need for pipe dope on compression fittings.

VALVE REPLACEMENT

The valves under your sink may be soldered on, screwed in, or attached with compression fittings. They might be angle stops, straight stops, stop-and-waste valves, gate valves, ball valves, or some other type of shut-off valve. To replace these valves, you must cut off all the water to the house. When the water is cut off, you can begin to replace your valves.

Most homes have either angle or straight stops under the sink. If the water pipes come up through the floor, they should be straight stops. When the water pipes come out of the wall, they should be angle stops. The first thing you must do is establish how the valves are attached to the water pipes.

If the valve has a large nut between it and the water pipe, it is a compression fitting. If a copper water pipe enters the stop and there is no nut, it is a soldered fitting. If there are threads visible going into the stop, it is screwed on. Compression fittings are the easiest to work with. If you have copper pipe, you can replace existing stops with compression stops.

Removing Existing Stops

In all cases, be sure the water is turned off at a point before the valve. Disconnect the supply tube from the stop. With screwed-on stops, turn them counterclockwise to remove them. This can be done with a wrench or t&g pliers. Grip the pipe coming out of the wall with pliers to keep it from turning with the valve. In a few turns, the valve will come off the threaded pipe.

With soldered valves, you have two options. You can cut the copper pipe to get the valve off, or you can heat the solder joint to remove the valve. If you have adequate pipe to work with, simply cut it to remove the valve. If you must remove the valve without cutting the pipe, you will need pliers and a torch. Placing the lighted torch under the solder joint will melt the joint. When the solder begins to drip, pull the valve off the pipe with your pliers. Be careful, as the pipe and the valve will be extremely hot. In this procedure, be very careful not to set your wall or cabinet on fire with the torch. Always wear safety glasses when working with hot solder.

With compression stops, you only need to loosen the large nut on the valve. This is done by turning the nut counterclockwise as you look at it from the pipe's direction. For example, looking under the sink at an angle stop, you will turn the nut clockwise as you are seeing it. The direction of turn will be counterclockwise if you run your hand up the pipe to the nut.

Replacing Valves
Replacing a screwed-on valve requires a wrench and pipe compound. Once you prepare the pipe threads with compound, screw the valve clockwise onto the threads until it is hand-tight. Then use your adjustable wrench to finish tightening the valve. Reinstall the supply tube and you are in business.

Installing solder-type stops requires more expertise and materials. When possible, use a compression stop instead of a solder stop. If you must solder the stop onto the pipe, you will need several items: sandpaper, flux, a flux brush, solder, a torch, a striker or match, and safety glasses. Sand the pipe until it shines. Apply the flux to the pipe with your brush. Sand the inside of the stop's fitting until it shines and apply flux. Slide the stop onto the pipe and position it.

With your safety glasses on, light the torch. Hold the torch under the fitting of the stop valve and occasionally place the solder on the top of the pipe, near the fitting. When the solder melts, it should run around and into the stop's fitting. This is how the solder joint is made. Don't touch the pipe or the stop; they are both very hot. Allow the pieces to cool down before inserting the supply tube. Again, beware of fire hazards with the torch.

Compression stops are simple to install. If you are replacing a compression stop, you can reuse the compression ring and nut. In a replacement situation, simply slide the new stop onto the pipe and connect the existing nut to the stop. Tighten it with your wrench while holding the stop with another wrench. When the connection is tight, you are all set.

If you are installing a compression stop to replace a solder stop, slide the nut on the pipe first. The threads should point toward the new valve location. Slide the compression sleeve (ferrule) on the pipe and then the valve. Tighten up the connection as described above and you are done. All that is left is to reinstall the supply tube.

Replacing Other Valves Under the Sink
While the style of the valve may be different, the principles of replacement are the same. By following the instructions above, you should be able to replace any valve under your sink.

Repairing the Valves Under the Sink
It is usually more cost-effective to replace bad valves than it is to repair them. If you choose to take your valves apart for repair, follow the instructions in Chapter 17 under Valves.

KITCHEN APPLIANCE PLUMBING

The three common kitchen appliances using plumbing are icemakers, garbage disposers, and dishwashers. When these units fail, the repair of interior parts falls into the category of appliance repair. Only the connection from the appliance to the plumbing system is considered a plumbing repair. Replacing these units is thought of as a plumber's job.

Icemakers

From a plumber's perspective, icemaker work is simple. The plumber's responsibility begins with the connection at the rear of the refrigerator. If your freezer has an icemaker, the refrigerator will have a ¼-inch threaded connection behind it. This is where the tubing for the water-feed to the icemaker is connected.

The tubing for an icemaker can be clear plastic or copper. In modern connections, the tubing attaches to the threads with a compression nut and ferrule. From the connection at the refrigerator, the tubing goes to a cold water pipe. This pipe may be under the kitchen sink or it could be at another location.

In today's plumbing, the tubing is connected to a self-piercing saddle valve by way of a compression nut and ferrule. To replace crimped or leaking tubing, you simply replace the old tubing and compression nuts and ferrules with new ones. Before doing this, you must close the valve at the saddle connection to cut off the water. If the saddle valve is leaking, it is also easy to replace.

Replacing the saddle valve. To replace the saddle valve, you must cut off the water to the pipe where the saddle valve is connected. Loosen the compression nut where the tubing enters the saddle valve. Next, loosen the screws holding the two pieces of the saddle valve together. The valve will come off the water pipe easily.

Place the rubber gasket on the saddle valve as shown in the instructions with the valve. Put the new saddle valve on the water pipe where you removed the old one. Turn the valve handle clockwise until it is engaged in the water pipe. Tighten the screws evenly to affix the saddle to the pipe. When the screws are tight, reconnect the tubing with the compression fitting.

Turn the valve handle clockwise until it will not turn any more. Cut the water back on to the pipe where the saddle is attached. Turn the saddle valve counterclockwise until it will no longer turn. At this point, there should be water flowing to the icemaker. If you wish to confirm your connection, you can loosen the compression nut at the tubing and saddle valve. When you loosen this nut, you should see water running around it. This confirms your connection. Tighten the compression nut and be proud of yourself. You just accomplished a job you would have had to pay a professional to do without this book and your talent.

Repairing saddle valves. Saddle valves are very inexpensive and are not worth repairing. If you have a bad saddle valve, replace it.

Dishwasher Plumbing

The water pipe for a dishwasher usually originates under the kitchen sink. The pipe is typically a ⅜-inch copper tubing. This ⅜-inch tube is not the same as supply tubes. Supply tubes are measured as ⅜ inch in *outside* diameter. Dishwasher tubing is measured by its *inside* diameter.

The water supply to the dishwasher should have its own shut-off valve. The valve may serve the hot water side of the kitchen faucet and the dishwasher. In this application, if the valve is turned off, the hot water to the faucet and the dishwasher is cut off. Another typical installation allows the dishwasher to have its own independent valve.

When the tubing leaves the valve and goes to the dishwasher, it joins a dishwasher ell. The dishwasher ell screws into the dishwasher connection and accepts the tubing with a compression fitting. This is as far as the plumber's responsibility extends for the dishwasher supply line. If you must replace the dishwasher ell, you need to apply pipe compound to the threads prior to installing it. The tubing is worked with like any other tubing and compression fittings.

Dishwasher drain hoses. The drains from dishwashers are usually rubber hoses. The hose coming out of the dishwasher has a diameter of ⅝ inch. The hose is connected at the base of the dishwasher to a male insert adapter with a stainless steel clamp or similar device.

The dishwasher drain hose should extend upward toward the countertop and into the base cabinet below the sink. At this point, the hose should connect to an air gap with a stainless steel clamp. Air gaps have two male insert fittings. The first accepts the drain from the dishwasher. The second fitting is sized to receive a hose with a diameter of 7/8 inch. This hose is also connected with a stainless steel clamp.

When the larger hose leaves the air gap, it connects to the drainage system below the sink. It may connect to a fitting on the side of a garbage disposer or to a wye tailpiece (Figure 5-23). Older homes may not have air gaps. In these homes, the smaller dishwasher hose will often be looped up under the counter and then run to the drain.

To connect these 5/8-inch hoses to disposers or wye tailpieces, you need a dishwasher adapter (Figure 5-24). The small hose is too small to fit over standard disposer connections and wye tailpieces without using the adapter. These adapters allow you to connect the small hose to the large drain receptor by using the adapter as a coupling. The connections to the adapter are made with stainless steel clamps. Dishwasher hose can be cut with a knife or a hacksaw. The main concern in working with this hose is to avoid leaving it kinked in the final installation.

Garbage Disposers

There are three basic repairs a homeowner can perform on garbage disposers. If your disposer fails to work at all, try pushing the reset button. This button is usually located on the bottom of the appliance. The button should be marked as a reset button. If you push the button and the appliance still fails to run, you probably need an electrician.

The second repair you may attempt is freeing a blocked impeller. If your disposer buzzes or hums, the cutters may be seized. Cut the power off to the unit before you do anything else. The sharp blades in a disposer don't know the difference between carrots and fingers. With the power off, place a broom handle in the mouth of the disposer.

When you feel the broom handle meeting resistance against the blades, apply moderate pressure to turn the blades. You may feel the blades break free from the obstruction. If the impeller blades refuse to

budge, don't force them. In most instances, the broom handle will free the blades. When you have done all you can do or you feel the blades free up, remove the broom handle. Turn the electrical power back on to the disposal. With a little luck, it will whir just as it should. If it doesn't, try the same procedure again. If it still doesn't work, call in a professional.

When your disposer springs a leak, you can fix it. Some disposers develop a leak where the disposer ell connects to the appliance. To replace this piece, you will need t&g pliers and a screwdriver. Loosen the slip-nut on the disposer ell where it enters the trap. Remove the screw in the retainer flange, and the disposer ell will come out. Place the new rubber washer from your repair parts kit over the new disposer ell. Slide the retainer flange over the ell and replace the screws you took out. Reconnect the slip-nut to the ell and everything should be fine.

To test this connection, fill the sink with water. Put the stopper in the sink and fill the bowl to the rim. Pull the stopper and watch your connection as the water drains. If you are going to have a leak, it will show up now. To test the connection effectively, you must fill the sink with water. Running water down the drain does not provide enough pressure against the connection to expose some leaks.

Replacing your garbage disposer. The plumbing aspect of this job is not too difficult, but the job requires working with electrical connections as well as plumbing. Since I am not a licensed electrician, I will not give you any advice on the wiring aspect of the job.

Removing the old disposer. Disconnect the slip-nut from the disposer ell. If necessary, remove the disposer ell from the appliance. Where the disposer meets the mounting ring there is a lock ring. This ring has tabs on it to assist in the removal process. Some rings can be turned by hand, and others require more force. You can use a hammer and a screwdriver to start turning the ring. Be advised: when the lock ring is loose, the disposer will fall to the floor or on top of you. Turn the lock ring counterclockwise to release the disposer from the mounting bracket.

The next step is to have the electrical wiring removed from the old disposer. When the old disposer is out of the way, have the electrical wiring connected to the new disposer. If you bought a replacement disposer of the same brand as the old one, you may be able to use the existing mounting bracket. If you can, place the disposer against the bracket and turn the lock ring clockwise.

When the disposer is locked in place, install the disposer ell. The last step is connecting the slip-nut to the ell. If you have to replace the mounting bracket, you must remove the old bracket. These brackets are ordinarily held in place by three threaded rods. When you place a screwdriver in the end of these rods and turn them counterclockwise, the bracket becomes loose. When you have relieved the pressure, you can easily remove the mounting bracket. Depending on the type of disposer you have, there may be a metal snap ring to be removed before the mounting bracket will come free.

After taking the old drain assembly out, remove any remaining putty from around the sink's drain hole. Follow the instructions that came with the new disposer to install the new drain assembly. Place putty around the flange of the drain and press it through the hole in the sink bowl. Under the sink, place the fiber washer over the part of the drain assembly protruding from the bottom of the sink. Next, place the upper portion of the mounting device over the drain.

In many cases, there will be a metal ring you must snap into a groove on the drain assembly to hold the mounting device in place. Tighten the threaded screws evenly until the flange is secure and the putty in the sink bowl has spread out. From this point on, follow the instructions given above for mounting disposers to existing brackets.

This concludes the section on kitchen plumbing. You will find that many products, such as faucets and garbage disposers, come with installation instructions. The best procedure you can follow when replacing kitchen components is to pay attention to how each part you are replacing is attached to its adjoining part. In this way, you will have a good idea of the relationship of the parts when you install or replace a new section.

6
Standard Bathroom Plumbing

Bathrooms contain a large portion of a home's plumbing. In a standard bathroom, there are a lavatory, a toilet, and a bathing unit. This chapter is going to explore all aspects of these standard bathroom fixtures and related plumbing. If your bathroom houses a spa, bidet, or other specialty fixture, you will find information on it in the next chapter.

THE LAVATORY

Lavatories can take many shapes and be made of many different materials. Your lavatory may be made of enameled steel, cast iron, plastic, fiberglass, or china, or it may be molded into a cultured marble top. There are other possibilities for the materials used in your lavatory, but these are the most common. The lavatory may be a drop-in, pedestal, wall-hung, or molded type. The requirements for repairing lavatories are about the same for all the different types.

Lavatory Faucets

Lavatories are usually drilled to accept faucets with a 4-inch center. This means there is 4 inches between the center of the faucet's water connections. Some lavatories are drilled for 8-inch center faucets. Most lavatory faucets have a one-piece body. Some more expensive faucets have three separate pieces to install in the mounting holes. These faucets have individual handles and a spout to mount on the lavatory (Figure 6-1). Old lavatories may

have two spigots with a separate handle and spout for the hot and cold water. The replacement and repair of most lavatory faucets is comparable to the kitchen faucet examples in the last chapter. The principles for repair and replacement are identical.

Under the Lavatory

Below the lavatory you should find a trap, two shut-off valves, two supply tubes, and the drain assembly. Except for the lavatory drain assembly, the workings under the lavatory are very similar to those of a kitchen sink. The procedures for working with the supply tubes and valves are the same as with the kitchen sink. The trap repair and replacement methods are the same, only the size may be different. Lavatories have a $1^{1}/_{4}$-inch drain instead of a $1^{1}/_{2}$-inch drain. The trap may be $1^{1}/_{2}$ inches with a reducing trap adapter connecting it to the $1^{1}/_{4}$-inch tailpiece.

The trap arm may be smaller than that of the kitchen example, but the repair and replacement methods are the same. The major difference between the lavatory and the kitchen sink is the drain assembly. Kitchen sinks use basket strainers and stoppers; lavatories normally use a pop-up drain assembly.

Lavatory Pop-Up Drain Assemblies

The pop-up assembly begins with the drain in your lavatory bowl (Figure 6-2). From there it extends below the bowl and connects to the trap. Near the middle of the assembly, there is a rod that extends out and toward the back of the lavatory in the direction of the faucet. This rod is connected to the lift rod that runs up through the faucet. When you pull up on the lift rod, the drain plug goes down into the assembly and prevents water from leaving the bowl. When you push down on the lift rod, the plug comes up and the bowl drains.

Adjusting Pop-Up Drains

Anyone with average ability can learn to adjust, repair, and replace pop-up drains. The first job is simple. If your pop-up plug will not stay up and causes the sink to drain slowly, you can fix it. After extended use, some plugs refuse to stay in the up position. To correct this, you must get under the lavatory. If you need any tools at all, pliers will do the job.

Looking at the drain assembly, you should see a rod extending toward the faucet. With your fingers, move this rod up and down. The rod controls the plug in your lavatory drain. Push the rod down until it will go no farther. Get out from under the lavatory and run water into the bowl. If the water drains properly, go back under the lavatory. Where the rod enters the drain assembly you will see a knurled, round fitting.

Tighten this fitting with your fingers or pliers, but don't get carried away. It does not take much to make these fittings tight. The fitting should be turned clockwise on the threads. When the fitting is snug, try moving the rod up and down again. It should be firm but not difficult to move. Now, from in front of the lavatory, pull up on the lift rod. The plug should go down and form a seal in the drain. Fill the lavatory with water and push the lift rod down. If the water drains well, you have corrected the problem. If it doesn't, you have another adjustment to make.

When this first attempt fails, you must look to the portion of the lift rod under the lavatory. As the pop-up rod coming out of the drain assembly extends toward the faucet, it is joined with a vertical piece by a retainer. As you look up the vertical piece, you

will see it is attached to the lift rod. This attachment is usually made with a set screw or nut. There are two possible scenarios for making this next adjustment.

Look at where the rod leaving the drain meets the vertical piece. If there are multiple holes on the vertical section where they meet, you may be able to make the adjustment there. Normally, there will be at least three holes that the horizontal rod may go through. If there is an open hole above the rod, you should be able to make the adjustment at this point. If the rod is in the highest hole, you will have to follow the instructions given later for the alternate method.

To make this adjustment, refer to the illustration of the pop-up assembly. You will see a small metal clip on the horizontal rod. Squeeze this clip with your fingers and slide it off the end of the rod. This procedure disconnects the rod from the vertical piece. Now, slide the metal clip back onto the rod, but only through one hole. Then place the vertical piece on the rod so the rod is in the next hole higher than its original position. Squeeze the clip and slide it completely on the rod. The vertical piece should be fairly straight all the way up to the faucet. If it is installed at an angle, the lift rod will be hard to operate.

Test your adjustment by pulling up the lift rod and filling the sink. When the sink is full, push down on the lift rod. The water should drain faster. If the sink still doesn't drain well, go to the alternate plan.

You may have to loosen the set screw or nut where the lift rod attaches to the vertical piece. You can do this with a screwdriver or an adjustable wrench depending on the type of mechanism you have. When the set screw is loose, push the lift rod up toward the faucet. Tighten the set screw and test the adjustment. You may have to work with it awhile, but with patience, you will accomplish your goal.

Repairing Pop-Up Drains

The illustrations show all the major parts of a pop-up drain. Normally, there are only two parts worth repairing. The first is the ball-and-rod assembly that comes out and connects to the lift rod. The other is the rubber washer where the assembly connects to the lavatory bowl.

Replacing the Ball-and-Rod Assembly

You must loosen the knurled fitting on the pop-up assembly to remove the ball-and-rod unit. When this fitting is loose, the ball-and-rod unit will pull out of the pop-up assembly. Note which hole the rod penetrates in the vertical piece connected to the lift rod before removing it. With the old ball-and-rod unit out, place the new one in the pop-up assembly. The ball portion will have a short rod extending into the pop-up assembly. This small rod may be placed through the pop-up plug or the plug may sit on top of it.

The plug will work better if the rod extends through it, but some people like to remove the plug to clean the sink. If the rod goes through the plug, you will not be able to remove the plug for routine cleaning of the bowl. The choice is yours; the drain will work either way.

When the ball is in place, replace the knurled fitting to hold the ball in the assembly. Most plumbers install pipe compound on the threads before putting the knurled fitting in place. This is an added precaution, but it should not be necessary. Install the retainer clip and vertical lift section next. Test your work by filling and draining the bowl. If you need to adjust the installation, refer to the examples earlier in the chapter.

When you think you are done, check for leaks around the knurled fitting. If it leaks, tighten the fitting until the leak stops. If the leak persists, remove the fitting, coat the threads with a sealant, and reinstall the fitting.

Replacing the Pop-Up Washer

There is a thick washer between the drain assembly and the lavatory bowl. To replace this washer, you will need t&g pliers, plumber's putty, and a thread sealant.

Turn the large nut near the washer counterclockwise. Loosen it until it is near the bottom of the threaded portion of the assembly. As it loosens, a metal washer should follow it down the threads. Remove the ball-and-rod assembly from the drain. When these are out of the way, loosen the slip-nut around the drain assembly as it enters the trap.

When the slip-nut is loose, push up on the drain assembly. It should extend upward into the lavatory bowl. Next, unscrew the finish trim fitting that you see when looking into the lavatory bowl. This is the piece where the plug enters the assembly and where the water drains out. Turn it counterclockwise to remove it. With it gone, pull the drain assembly out of the bowl from below. Remove the old washer and install the new one. The flat side should face the trap, and the beveled side should face the bowl.

Slide the washer down on the drain enough to allow the threads to extend back up into the lavatory bowl. Place putty around the underneath of the finish trim piece you removed earlier. Apply thread sealant to the fine threads and push the assembly back up through the hole in the lavatory. Screw the trim piece onto the threads. Push the entire unit down until the putty spreads out.

Go back underneath the lavatory and tighten the large nut. When it is tight, reconnect the ball-and-rod assembly. Then reconnect the trap to the tailpiece. When all your connections are reassembled, you are ready to test for leaks. Fill the bowl to capacity with water and let it all drain at once. Wipe all connections and parts with a paper towel to identify any leaks. If you have leaks, tighten the fittings at the problem area.

The Lavatory Mystery Leak

If your lavatory will not hold water, but doesn't appear to be leaking, you have a putty problem. The putty under the trim piece of the pop-up in the lavatory bowl needs to be replaced. Without putty, water will run under the trim piece and into the drain assembly. The water will drain out of the bowl and down the drain. This type of leak can drive you crazy trying to find it.

The solution to this is to install new putty under the finish trim of the pop-up assembly where the plug

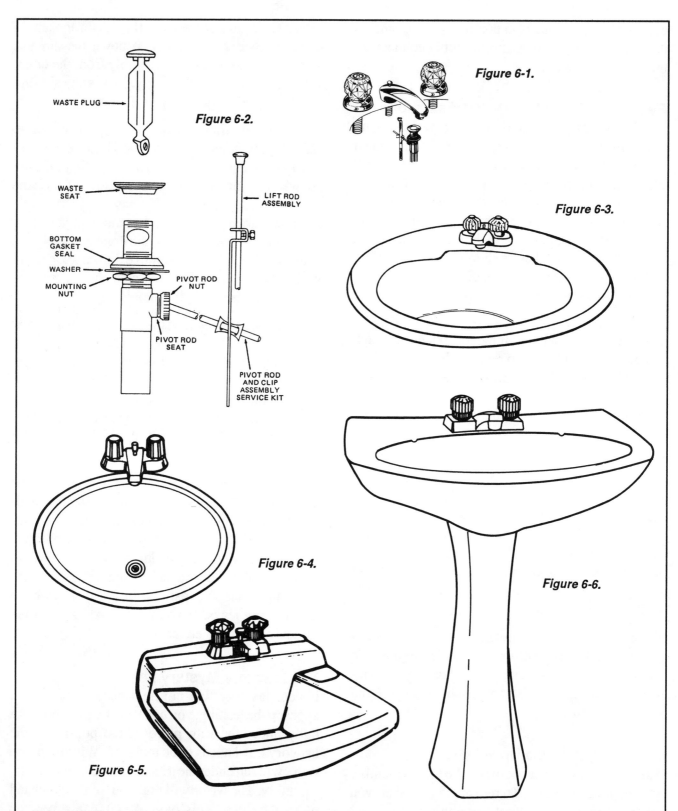

Figure 6-1. Three-piece lavatory faucet. Courtesy of Moen, Inc. **Figure 6-2.** Pop-up lavatory waste assembly. Courtesy of Moen, Inc. **Figure 6-3.** Self-rimming lavatory. Courtesy of Universal-Rundle. **Figure 6-4.** Rimmed lavatory. Courtesy of Universal-Rundle. **Figure 6-5.** Wall-hung lavatory. Courtesy of Universal-Rundle. **Figure 6-6.** Pedestal lavatory. Courtesy of Universal-Rundle.

sits. Follow the same steps taken for replacing the pop-up washer to gain access to the trim piece. Instead of replacing the washer, just put new putty around the trim piece and replace the assembly. While you have the assembly apart, inspect the washer. It may be worthwhile to replace it while you have everything torn apart.

Replacing Your Lavatory

There are four basic types of common lavatories. They are: *self-rimming* (Figure 6-3), *rimmed* (Figure 6-4), *wall-hung* (Figure 6-5), and *pedestal lavatories* (Figure 6-6). Pedestal lavatories are discussed in Chapter 7.

To replace your lavatory, turn off the water to the supply tubes. Disconnect all waste and water connections from the bowl. If you have doubts on this procedure, review the early portion of this chapter and the sections in Chapter 5 on replacing a kitchen sink. When all connections are removed, replace the bowl according to the following instructions.

Replacing self-rimming lavatories. Self-rimming lavatories simply sit in the hole of the counter. You may have to break the caulking seal around the edges of the bowl before it will lift out of the hole. This can be done with a utility knife. It is usually an easy task to remove self-rimming lavatories. Pushing upward on the bowl from the bottom, under the counter, is all that is required for removal.

Before installing the new bowl, clean the counter surface around the hole. When the surface area is clean, place the bowl in the hole. Align it in the position you want and reconnect the waste and water fittings. Most plumbers install the new faucet and pop-up assembly before putting the bowl into the counter. This procedure enables you to make the mounting connections standing up. If you install the bowl before mounting the faucet and pop-up, you will have to work from under the counter to install them.

When the bowl is set and connected to the waste and water pipes, check the alignment of the bowl. If the bowl is sitting where you want it, the next step is to seal around the bowl. You can use any of the various

caulking compounds available to seal the bowl. If you do not caulk around the bowl, water splashed on the counter may run under the bowl and leak into the cabinet. This water could ruin your cabinet or the inner portion of the countertop.

Run a line of caulking around the perimeter of the lavatory bowl. Use your finger to trace over the caulking. The pressure from your finger will push the caulking into the crack between the bowl and the counter. When the crack is filled, rub a wet paper towel over the caulked area to clean up the excess caulking.

Replacing rimmed lavatories. The only big difference between a rimmed bowl and a self-rimming bowl is the rim. The rim is a metal ring surrounding the bowl. The ring fits in the hole and uses clips to hold the bowl in place. The principle is very much like that used with stainless steel kitchen sinks.

To accomplish this replacement, follow the instructions given to this point but make the following changes. From under the sink, loosen the sink clips, but be warned, when the clips are loose, the bowl will drop. If you're not careful, you could have a lavatory bowl come crashing down on top of you. When the bowl is removed, remove the ring. This can be done with your hands or a pair of pliers if the ring is securely attached to the counter. You may have to cut the caulking under the ring before removing it.

Instead of placing the bowl in the counter hole, put the ring in the hole. This job is much easier if you have someone to help you. With the ring in place, you must hold the lavatory bowl in place from under the cabinet. While you are holding the bowl, you must install the clips. This can be a juggling job if you are working alone.

As you tighten the clips, check the alignment of the bowl. It is easy for the bowl to become turned as you install the clips. When the bowl is right, tighten the clips. From this point on, the procedure is the same as in the instructions given above.

In deciding to replace a rimmed bowl, you may want to consider replacing it with a self-rimming

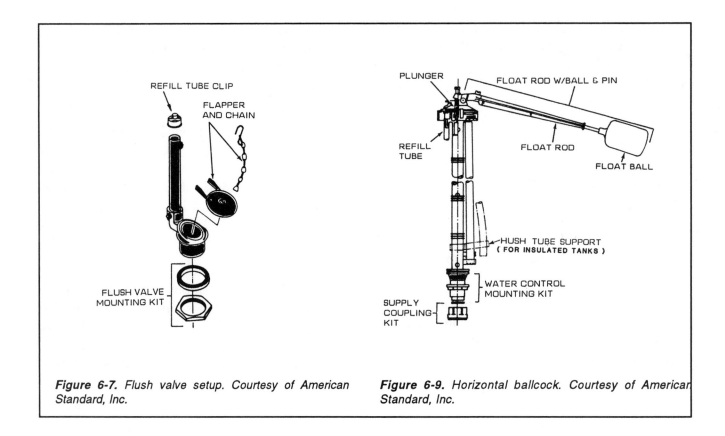

Figure 6-7. Flush valve setup. Courtesy of American Standard, Inc.

Figure 6-9. Horizontal ballcock. Courtesy of American Standard, Inc.

bowl. Self-rimming bowls are easier to install and easier to clean. If you want to switch bowl types, measure the hole in the cabinet and find a self-rimming bowl that will fit and cover the hole.

Replacing wall-hung lavatories. Disconnect all waste and water connections. Grasp the sides of the lavatory and lift it upward. The bowl should come loose without much effort. Wall-hung lavatories sit on brackets and simply lift off for removal. Some older lavatories have bolts under them, securing the bowl to the wall. If your bowl doesn't lift off easily, check for these bolts under the bowl where it meets the wall.

In most cases, the existing wall bracket can be used with the new lavatory. Mount your new faucet and drain assembly first. Then position the lavatory on the wall bracket and push downward to seat the bowl on the bracket. If the mounting doesn't go well, check to see if you must install a new bracket. It is rare that you need to use a new bracket, but there may be times when you must.

Installing the bracket is a matter of removing the old bracket and screwing the new one to the wall. There should be wood backing behind the finished wall to accept the mounting screws of the bracket. Even wall-hung lavatories with legs must first be mounted to a wall bracket.

When you have the new bowl securely on the bracket, put a level on the top of the bowl pointing from left to right. With the heel of your hand, tap the bowl as needed to level it. When the bowl is level, make your connections to the waste and water fittings. A word of advice: when making your connections, don't push up on the lavatory. If the bowl is pushed up off the bracket, it will fall. This can leave you with a broken lavatory or a sore head.

Lavatory Legs

To remove lavatory legs, turn the legs to release the pressure on them. Clockwise is the normal direction for turning the legs, but some may need to be turned counterclockwise. As you turn the leg, you should

Figure 6-8. Vertical ballcock.

feel tension if you are turning it the wrong way. When the leg is loose, swing it out toward you and it should drop out from under the bowl.

To install legs, follow the directions supplied with the new legs. There should be a threaded rod to screw into the top of the leg. This rod fits into the small holes under the two front corners of the lavatory. When the rods are in place, turn the legs to give the desired tension and support.

TOILET TROUBLES

The toilet can present some interesting challenges for the novice plumber. At first look there doesn't seem to be much that could go wrong with a toilet, but looks can be deceiving. Standard toilets can produce a multitude of problems for the weekend plumber. This section of the chapter outlines those problems and their solutions.

The Running Toilet

If your toilet continues to run after a flush, it is costing you money. To correct the problem, you must determine what is causing the malfunction. It could be in the ballcock or the flush valve. The first place to look is inside the toilet tank.

Look at the flush valve assembly (Figure 6-7). Is there water running over the refill tube? If water is filling the tank and running down the refill tube, your problem is in the ballcock. If the water level is below the refill tube and still leaking into the toilet bowl, the problem is in the flush valve assembly.

Adjusting the Ballcock

The most common cause of running toilets is the ballcock. There are two common styles of ballcocks. One utilizes a *vertical float* (Figure 6-8), and the other uses a *horizontal float rod and ball* (Figure 6-9). Both types can be adjusted to control the tank's water level.

With the horizontal style, all you have to do is bend the float rod. Bending it so the float is lower in the tank will cause less water to come into the tank. Bending it so the ball is in a higher position allows additional water to fill the tank. In the case of a running toilet, you would bend it downward.

Place the palm of one hand under the rod and use your other hand to bend it downward from the end of the rod where the ball is attached. This is an extremely simple plumbing adjustment. With this type of ballcock, the float sometimes becomes stuck against the side of the toilet tank. This can also cause the water to continue running after a flush.

Float rods are threaded and screw into the ballcock. When this connection is loose, the float ball can become turned by the water until it is lodged against the tank. When you look into your tank, be sure the float is not touching the tank. If it is, gently bend the float rod to free the float, or tighten the connection between the float rod and the ballcock to eliminate the problem.

With the vertical type of ballcock, the float moves up and down on the shaft of the ballcock. This style does not become stuck against the tank. To adjust this style, there is a small metal rod and a metal clip running parallel to the ballcock shaft. By squeezing the metal clip, you can move the float up or down to allow for water adjustment.

Some of the vertical ballcocks can be rotated on the shaft for water level adjustment. There should not be a circumstance when this is needed in existing plumbing. When the toilet has functioned properly in the past, only minor adjustments should be needed. The minor adjustments are done with the metal clip.

Replacing Float Balls

Replacing the float on a ballcock is extremely simple. There are no tools required for this task. Unscrew the existing ball from the float rod and screw the new ball on in its place. You can substitute Styrofoam®, plastic, and metal float balls for other types without a problem.

Replacing a Ballcock

To handle this job, you will need a few tools, a sponge, and a small pan or pail to catch water in. If things go as planned, the only tool you will need is an adjustable wrench. With these items and a new ballcock, you are ready to do the job. It is good insurance to have a replacement closet supply and a new 3/8-inch ferrule on hand.

This is a simple job when things go right. First, turn off the water to the toilet. This is normally done at the valve directly under the toilet tank. Flush the toilet to be sure the valve is holding and to empty most of the tank's water. Use the sponge to remove the water remaining in the tank after the flush. When you remove the old ballcock, any water left in the tank will run out on the floor.

With the water off and the tank empty, loosen the ballcock supply nut with your wrench (Figure 6-10). When this nut is loose, it will fall down onto the cutoff valve. Gently move the closet supply tube away from the existing ballcock. Remove the nut pressing against the base of the tank to remove the ballcock. When the retaining nut is off, the ballcock will lift out of the tank.

Place the new washer on the new ballcock and push the assembly through the hole in the tank. Install the new retainer nut to hold the ballcock in place, but don't over-tighten it. It is easier to snug it up later than to repair cracked china. With this done, gently put the existing closet supply tube back in place. Normally, the fit will be fine, and you can reuse the old supply tube and the old supply tube nut.

If the supply tube is crimped or doesn't mate up, you will have to replace it. You learned how to do this in Chapter 5 with sink supplies. Use the same techniques with the closet supply tube.

Adjusting the Flapper or Flush Ball

When your running toilet is being caused by the flush valve, you should look to the flapper or flush ball first. The flush valve will normally have either a *rubber flapper* (Figure 6-11) or a *ball-type device* (Figure 6-12) that covers the hole between the tank and the bowl of the toilet where the water runs. These devices frequently become defective and cause a toilet to run constantly.

Occasionally, the chain running from the toilet handle to the flapper will get tangled. Sometimes it will wrap around the flapper, keeping the flapper from seating on the flush valve. If you cannot see any evidence of these two conditions, examine the flapper. Normally, if the flapper is sitting on the flush valve but still leaking, you must replace it. This is an easy operation explained under the next subheading.

Tank balls are attached to a guide wire. There is very little opportunity to adjust tank balls. If your tank ball is worn, you may see a hole in it. A damaged ball can appear fine. If the ball is seated on the flush valve and leaking, replace it.

Replacing Flappers and Tank Balls

Replacing either of these items is easy. This job can be done with the water to the toilet on, but you will not waste as much water if you cut the supply valve off. First we will look at how to replace the flapper. Your flapper should have either two tabs (Figure 6-13) or a round circle (Figure 6-14) at the end of it.

The ones with tabs attach to two points on the side of the refill tube. The ones with a circle slide down over the refill tube. There will be a chain or rubber line running up from the flapper to the toilet handle.

The replacement only requires you to remove the old flapper at these points and install the new one. This is a very simple job, and it normally requires no tools. The circle-type flappers are fairly universal. They can usually be used to replace either type of flapper.

Tank balls are screwed onto the guide wire. To remove the old one, you simply unscrew it. Installing the new one is just as simple. All you have to do is screw it onto the guide wire. In extremely old toilets, you may encounter different types of tank balls. Generally, if the tank ball does not screw onto a guide wire, you will have to replace the flush valve.

Replacing Lift Wires

These are the wires connecting the tank ball to the toilet handle (Figure 6-15). The lift wire is typically bent at one end and inserted through a hole in the handle. The other end has a factory bend in it for the other wire to pass through. The first wire has a threaded end for the tank ball to attach to and a factory bend to retain it in the lift wire.

To replace these wires, you must unscrew the tank ball and remove the lift wire from the handle. That is all there is to removing the wires. To replace the wires, put the first wire through the factory bend in the lift wire and feed it through the guide coming off the refill tube. Screw the tank ball onto the first wire. Bend the end of the lift wire so it will penetrate one of the holes in the toilet handle. Then bend the lift wire passing through the handle down, so that it will not pull loose from the handle.

Test your installation by flushing the toilet. If the tank ball seals and the water stops, you have accomplished your task. If the tank ball does not seal, you need to adjust the wires. This usually means putting the lift wire through a different hole in the handle or reworking your bend to allow more length to the lift wire.

In some cases, you will have to adjust the guide coming off the refill tube. This can usually be done without tools by grasping the guide and turning it with your hand. It is important for the guide to be lined up directly over the flush hole in the tank.

Replacing the Refill Tube

Metal refill tubes occasionally develop holes. When this happens, the water in the tank runs constantly into the toilet bowl. If you notice water running into the bowl at odd times, check the refill tube for defects.

The refill tube is usually easy to replace, if it is replaceable. Some flush valves use plastic fill tubes that are made into the flush valve and are not replaceable. Others use brass tubes that may be replaced. If your refill tube is replaceable, you replace it by unscrewing it from the flush valve. You should be able to do this with your hands. Once the old one is out, screw in the new one.

Replacing the Flush Valve

When flush valves are beyond repair, they must be replaced. This can be a difficult and sometimes dangerous job. Old flush valve nuts seize up and can be very demanding to remove. In exerting pressure on the nut, the tank may crack or shatter. If the tank shatters, pieces of china will be flying and may get in your eyes. Safety glasses are a must for this job.

In addition to safety glasses, you will need tools. The job requires a pipe wrench in most instances. An adjustable wrench and a common screwdriver are also needed. When you attempt to change a flush valve, you are working with almost all the parts of the toilet tank.

You will have to work with the closet supply tube, the tank-to-bowl bolts (these can be a problem too), the tank-to-bowl gasket, and the flush valve assembly. On paper, this job appears simple enough, but in reality it can get very complicated. Unless you are adept at repairs, this might be a good time to call a professional. If you are convinced you can do the job, this is how it's done.

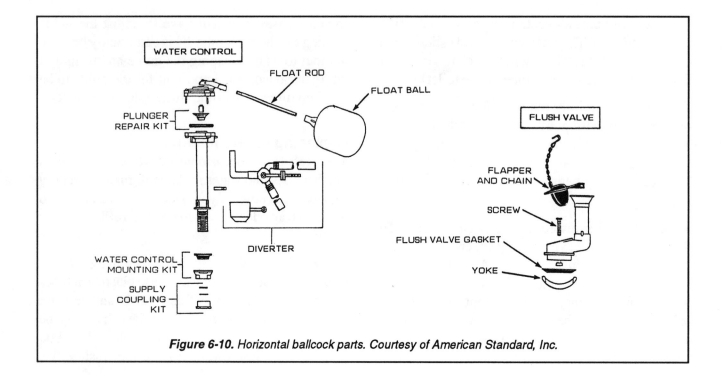

Figure 6-10. Horizontal ballcock parts. Courtesy of American Standard, Inc.

Cut the water off and flush the toilet. Sponge all the water out of the tank. Replacing a flush valve requires that you remove the tank from the bowl. Disconnect the closet supply from the ballcock. Put a small adjustable wrench on the nuts holding the tank to the bowl. Use a common screwdriver to turn the tank-to-bowl bolts inside the tank.

The tank-to-bowl bolts may have deteriorated and might not cooperate. If the head of the bolt dissolves as you try to turn it, you will have to cut the bolts off with a hacksaw. If you must saw the bolts off, be careful. The saw blade will scratch the china, and there will be minimal room to work. The best approach is to use a hacksaw blade in a jab saw.

Once the tank-to-bowl bolts are out, there is nothing to hold the tank onto the bowl. This job goes best for the weekend plumber when someone is available to hold the tank during this operation. With the bolts gone, lift the tank off the bowl and place it on its back, on the floor. With most toilets, you will see a small, thin black gasket. This is the tank-to-bowl gasket. Remove this gasket if it appears worn or damaged. You may need a replacement for it before putting the tank back on the bowl.

Looking at the bottom of the tank, you will see the large threaded portion of the flush valve. It should have a large sponge or rubber gasket fitted over the threads. Remove this gasket and inspect it. You will probably need to replace this gasket before reinstalling the tank. These gaskets and the tank-to-bowl bolts do not come with the new flush valve. Each of these items must be purchased separately.

With the large gasket off, you should see a big nut pressing against the bottom of the tank. In some cases, the nut can be removed with t&g pliers, but a pipe wrench will probably be needed. Before trying to loosen this nut, be sure you are wearing safety glasses. Attempt to turn the nut counterclockwise. If the flush valve is plastic, the nut should turn easily. If it is brass, you may be in for a struggle.

There is a high risk of breaking your tank if the nut does not turn freely. Any torque applied against the tank can break it. If the nut refuses to turn, try some spray lubricant on the threads. After spraying the threads, let them sit for five minutes. Try turning the nut again. If it still does not turn, seriously consider calling a professional plumber.

Figure 6-11. Flapper. Model: Andrew E. Wallace.

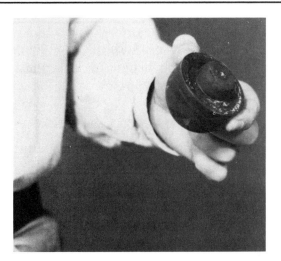

Figure 6-12. Tank ball. Model: Andrew E. Wallace.

Professionals know how to deal with these stubborn fittings. They can often be loosened by applying heat to the threads, but this method is too dangerous for the average person. The heat can cause the china to explode and become tiny but sharp missiles. *Do not attempt heating the flush valve to remove it!*

Assuming you were able to get the valve out, this is how you install the new one. The flush valve will have a tapered gasket that came with it. Slide this gasket over the threads until it touches the base of the flush valve. Install it with the flat side up toward the flush valve. This will leave the tapered part pointing down toward the toilet bowl when installed.

Insert the flush valve through the hole in the bottom of the tank. Install the large nut and snug it up. This nut must be tight, but too much pressure will crack the tank, so be careful. You will not be able to tighten this nut after the final installation without removing the tank again. Next, place the rubber or sponge gasket over the threads. This should be a new gasket, but you might be able to use the old one. If necessary, replace the thin tank-to-bowl gasket.

You are now ready to reinstall the tank. Set the tank on the bowl with the flush valve lined up over the flush hole. Use new closet bolts when you reinstall the tank if you have them. Place the rubber washer

that came with the bolts over the threads until they seat against the head of the bolt. Insert the bolts through the holes in the tank and bowl. Put the thin metal washers and nuts on the bolts. Tighten the nuts until the bowl is secure on the tank. You should tighten these bolts evenly. Don't tighten an individual bolt and then move on to another. Tighten each of the bolts a little at a time to avoid stress on the china. Don't over-tighten these bolts; you can adjust them later.

Reconnect the supply tube to the toilet and turn the water on. Flush the toilet to test your connections. Look for leaks at the supply tube, at the tank-to-bowl bolts, and where the tank meets the bowl. If the supply tube connection leaks, tighten the supply nut. If the tank-to-bowl bolts leak, tighten them a little at a time until the leak stops. If you have a leak where the tank meets the bowl, you may have to remove the tank.

Tightening the tank-to-bowl bolts may compress the flush-hole gasket and solve your problem. You may have to tighten the large nut on the flush valve to stop the leak. The point to remember is that too much tightening will result in a broken tank.

The exception to this procedure involves wall-hung toilet tanks. With wall-hung toilet tanks, you do not have to remove the tank from the bowl. Wall-hung tanks use a flush ell to connect the tank to the bowl. With these toilets, you simply loosen the large nut where the flush ell meets the flush valve. With this connection removed, the remainder of the replacement procedure is the same.

Tank-to-Bowl Bolts and Gaskets

You can refer to the instructions on replacing flush valves to find information on replacing tank-to-bowl bolts and gaskets.

Replacing Flush Ell Washers

Wall-hung toilet tanks connect the tank to the bowl with a flush ell. The two large nuts on the flush ell use slip-nut washers to maintain their seal. If you have a leak around these nuts, replacing the washers should cure the leak.

To replace the washers, cut off the water and flush the toilet. Loosen the nuts and remove the ell from the fitting at the tank or bowl. Slide the old washer off and the new washer on. Replace the ell and tighten the nuts. That is all there is to it if the ell is in good shape.

Flush ells tend to deteriorate with age. When you attempt to work with the ell, it may crack open from the movement if it has lost its structural integrity. When this happens, you must replace it.

Replacing Flush Ells

To replace your flush ell, cut off the water and flush the toilet. Loosen the large nuts and remove the ell from the two connection points. The new flush ell will normally need to be cut to size. You can cut a flush ell with a hacksaw or large tubing cutters. When you make your cut, try to use the old ell as a template. By using the old ell as a template, you will have an accurate assessment of the proper length.

If the old ell is in pieces, you will have to experiment. Hold the ell in the general position and mark it for your cuts based on the estimated length. One of the easiest ways to estimate the length is to put the ell into the tank hole. Then measure from the center of the vertical section to the desired location in the bowl. This will give you a good idea of the length

needed on the horizontal section. You can place the new ell in the bowl and measure from its center point up to the desired location in the tank to estimate the other cut.

When cutting the ell, leave it a little long on the first cut. It is much easier to cut more off than it is to put the cut pieces back together if your estimate is a little off. When you have the ell cut to size, slide the nuts onto the ell with the threads facing the tank and the bowl. Slide the washers onto the ell and insert the ell into the tank and bowl. With the ell in place, tighten the nuts.

This procedure can be frustrating. It can be very trying to get both ends of the ell into the tank and the bowl at the same time. You may want to call a pro to handle this job for you.

Fighting Tank Condensation

Toilet tanks can produce enough condensation to rot your bathroom floor. Condensation is a result of cold water in the tank and warm room temperature. To eliminate the sweating on your tank, there are a few options.

You can install Styrofoam® panels inside the tank to insulate it, but this rarely works efficiently. You can install a tank cover to collect the moisture and change it when the cover is saturated. The only sure way to cure the problem is the installation of a mixing valve. This valve allows you to mix hot and cold water before it enters the tank. By controlling the temperature of the tank water, you can stop the condensation.

To install a mixing valve, you must have access to a hot water pipe. The hot water pipe must run to the mixing valve where it is met with a cold water pipe. From the mixing valve, a single pipe carries the mixed water to the toilet. This can be a very complicated job and is best left to professionals.

Replacing the Toilet Handle and Trip Lever

Toilet handles are secured to the tank with a nut. New toilets have handles with the trip lever molded as a part of the handle (Figure 6-16). Older toilets have the handle with the trip lever a separate part, attached with a set screw.

Remove the handle by removing the nut on the inside of the tank. The nut will slide down the trip lever of a one-piece unit. In the two-piece design, loosen the set screw to remove the trip lever, and the nut will come off the handle. Reverse the procedure to install the new handle and trip lever.

Replacing the Wax Seal and Closet Bolts

This job requires the removal of the toilet from the floor. Cut the water off and flush the toilet. Use a plunger to force the remaining water in the bowl down the drain. If you don't, the bowl water will spill on the floor when the toilet is lifted. Disconnect the supply tube from the tank. Remove the covers from the closet bolts.

The toilet may be secured to the flange or floor with bolts or lag screws. Most toilets are held in place with bolts. To remove these bolts, turn the nuts counterclockwise. If you have lag bolts, turn them counterclockwise to remove them. When the nuts are removed or the lag bolts are gone, you can lift the toilet. Before removing the toilet, prepare a spot to set it temporarily. Place cardboard or paper on the floor to set the toilet on. There will be old wax on the bottom of the toilet that may get on your floor without this type of protection.

When you are ready, lift the toilet off the floor and put it aside. If the weight of the toilet is a consideration for you, the tank can be removed to lighten the load. When the toilet is moved, you should see a closet flange. The closet flange will have a round disk that surrounds the drain pipe. The disk will have slotted holes in it for closet bolts to sit in. The flange should be anchored to the floor with screws. If you don't have a closet flange, you should call a plumber to install one for you. If you elect not to have a flange installed, you can proceed, but I recommend having a flange installed.

Remove any old wax from the flange, floor, and around the pipe opening. This can be done easily with a putty knife. Remove the existing closet bolts from the flange. They usually lift right out when

aligned with the slot in the flange. Install the new closet bolts by putting them through the large slot in the flange and sliding them to line up with the center of the pipe. If you are working without a flange, there will be no bolts to install at this time; you will use lag bolts later.

Unwrap and place the wax ring on the flange so that it lines up with the pipe opening. The wax ring should be warm to work well. If you are doing this job in an unheated, cold room, warm the ring before installing it. (It can be warmed under a light bulb, on the dash of a car with the heater/defroster on, or carefully with a torch.) Without a flange, place the ring over the pipe and on the floor. This area is now ready to accept the toilet. Tilt the toilet and remove the existing wax. Next, lift the toilet and set it down on the wax ring. The closet bolts should line up with the holes in the base of the toilet. When the toilet is sitting on the wax ring, push down on the toilet to compress the wax.

Adjust the toilet to align it with the back wall by turning it as needed on the wax. When it is where you want it, install the nuts on the closet bolts. If you are working without a flange, screw the lag bolts into the floor through the holes in the toilet's base. Tighten these nuts or bolts until the toilet is firmly secure, but don't over-tighten.

Reconnect the closet supply to the tank and cut the water on. To test your work, flush the toilet several times. If water does not seep out from under the toilet, the connection is good. If water does seep out, you will have to start over again with a new wax ring. If the toilet flushes slowly or stops up soon after this job, you may have spread the wax out over the pipe opening. If the wax was not centered or shifted during the exchange, it may be blocking the drain.

Replacing the Toilet

Toilets with the tank mounted on the wall can be replaced with conventional toilets. To replace a conventional toilet, follow these instructions. Cut off the water and remove the old toilet using the information given throughout this chapter.

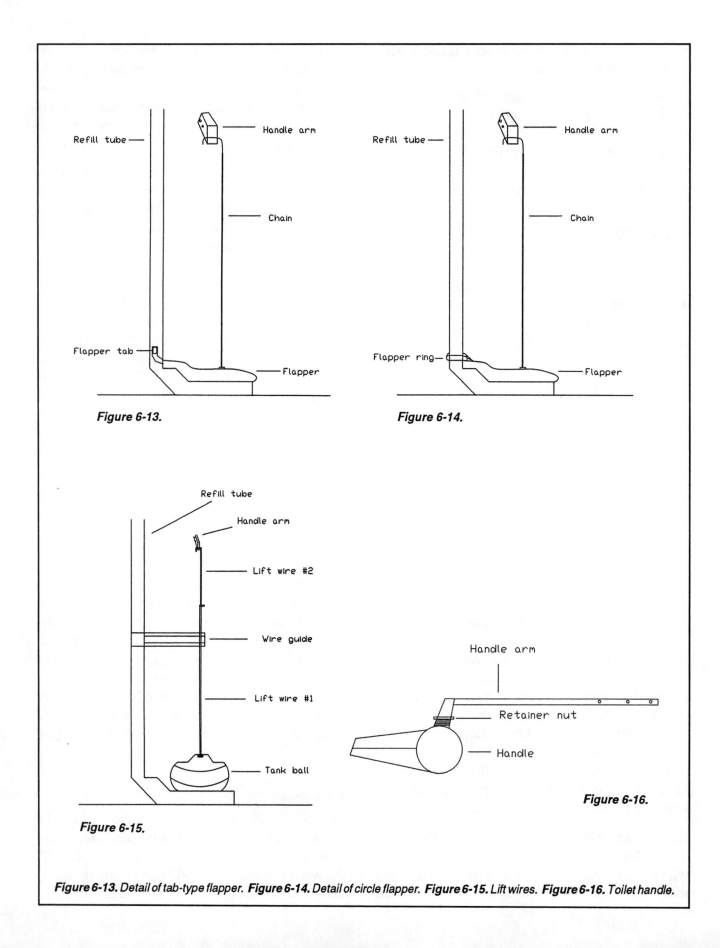

Figure 6-13. Detail of tab-type flapper. **Figure 6-14.** Detail of circle flapper. **Figure 6-15.** Lift wires. **Figure 6-16.** Toilet handle.

Unpack your new toilet combination and follow the factory instructions. The first step is to set the bowl. This is done with the same procedures described in replacing wax rings. Clean the flange and remove the old wax and bolts. Install the new wax and bolts. Set the bowl and bolt it down. You can measure from the back wall to the seat holes to achieve proper alignment.

The new tank should have all the parts installed. Place the sponge or rubber gasket over the flush valve. Install the tank-to-bowl gasket if the toilet is designed to have one. Not all toilets use these gaskets, so refer to the manufacturer's instructions to see if one is included. Put the rubber washers on the tank-to-bowl bolts. Set the tank on the bowl and install the tank-to-bowl bolts.

Reconnect the supply tube, turn on the water, and test your installation. A professional plumber can do this job in less than thirty minutes if things go well.

BATHTUB AND SHOWER PROCEDURES

Bathtubs and showers share many characteristics. Their faucet bodies are concealed in the wall. Their drains are frequently concealed between the bathroom floor and a ceiling below them. Many bathtubs are equipped with shower heads. In this section, you will learn about both bathtubs and showers.

Replacing Your Shower Head and Arm
This is a simple job. Place an adjustable wrench on the flat spots where your shower head meets the shower arm (Figure 6-17). Turn the head counterclockwise to remove it. You may have to hold the shower arm to keep it from turning with the shower head. In just a few turns, the shower head will come off.

To install a new head, put pipe compound on the threads of the shower arm. Screw the new head onto the arm with your hand. When it is hand-tight, tighten the head with your wrench. That's all there is to it.

If you want to replace the arm, follow the same procedure, but don't hold the arm to keep it from turning. If it does not turn with the head, you can use pliers to remove the old arm. To install the new arm, slide the escutcheon (the ring covering the pipe's entry into the wall) onto the long section of the arm. Apply pipe compound on the threads at both ends of the arm. Screw the shower head onto the threads of the short bend in the arm. Put the arm into the wall and screw it into the female threads where the old one was removed.

Put your wrench on the flat spots and turn the wrench clockwise. As you tighten the head, the arm will turn and tighten with it. Do not use pliers on the new arm — they will scratch the finish. If you must use pliers, keep them on the arm close to the wall so the escutcheon will hide the scars.

Repairing Leaking Faucets in the Tub or Shower
This job requires the use of tub wrenches (Figure 6-18) with most faucets having two and three handles. Single-handle faucets don't require tub wrenches. In addition to tub wrenches, you will need an adjustable wrench and screwdrivers.

Single-Handle Tub or Shower Faucets
Repairing these faucets is very similar to the examples given for kitchen faucets in the last chapter. The two basic types are cartridge and ball faucets. Before repairing these faucets, you must cut off the water to the faucets.

Some tubs and showers have cut-offs near the faucets. These valves are in the wall and are normally accessible through a removable panel or door. Frequently, these access areas are in closets that share a common wall with the bathing unit. If you do not find an access panel, you can look below the bathing unit if you have a crawlspace or basement. Sometimes the valves are below the floor in these accessible places. Not all bathing units have individual cut-off valves. If you are unable to locate valves for the faucets, cut off the water at the main shut-off to the house.

Replacing the Faucet Cartridge

Once the water is off, you are ready to begin (Figure 6-19). Remove the cap covering the screw to the faucet handle. Remove the screw and the handle. Remove the stem escutcheon. Be positive the water is off and the water pressure is relieved before going any further. Next, remove the clip holding the cartridge in place. Grip the end of the cartridge with your pliers and pull it out. Push in the new cartridge and reverse the take-down procedure to put the faucet back together. Test your repair.

Replacing Ball Assemblies

Cut off the water to the faucet. Remove the handle and unscrew the retainer nut from the face of the faucet. Remove the ball assembly and install the replacement. Put the faucet back together and test your repair.

Repairing Multi-Handle Faucets

Whether your faucet has two or three handles, the repair process is the same. Before doing any of the repairs, cut off the water. In all cases, remove the handles and trim escutcheons. The normal repairs include seat replacement, washer replacement, "O" ring replacement, and the replacement of packing material.

Removing the Stem

For all these repairs, you must remove the faucet stem from the body (Figure 6-20). After the handle and escutcheon are removed, you will be looking at the body and the stem. The stem is held into the body by threads. The threaded portion is recessed in the wall. This makes it very difficult to remove the stems without tub wrenches.

Slide the proper size tub wrench over the stem and onto the flat section near the body. Turn the wrench counterclockwise to remove the stem. This may take some force if the faucets are old. When the stem is out, you are ready to do your repair work.

Replacing Washers

The washer is attached to the bottom of the stem with a screw. Remove the screw, and the washer, or what's left of it, will come right off. Choose a replacement washer and put it in the brass housing at the bottom of the stem. Replace the screw and reinstall the stem. Test your work to see if the drip has stopped. If the faucet still drips, try a larger washer.

Replacing "O" Rings

The "O" rings are found on the outside of the stem. You simply remove the existing rings and replace them with new ones. This will stop water leaking around the stem.

Replacing Packing Material

Packing material is a greasy string found under the packing nut. To replace the packing, remove the packing nut and old packing. Wrap new packing material around the stem and replace the packing nut. This will eliminate water leaking past the packing nut.

Multi-Handle Faucet Cartridges

The faucets using these cartridges come in many variations. Normally, the cartridge is held into the body with a retaining nut. When the retainer is removed, the cartridge should come out (Figure 6-21). You install the new cartridge and reinstall the retainer.

Replacing Tub and Shower Faucets

The replacement of tub and shower faucets can be very involved and potentially dangerous. It requires the use of a lighted torch on the inside of your home's wall. The risk of fire is great when the job is done by someone other than a professional. Unless you are an advanced weekend plumber, call a professional for this job.

The first step in this job is cutting off the water to the pipes feeding the faucet. Drain the water from a lower point to clear the pipes of water. The second step is gaining access to the pipe and faucet body. If you don't have an access panel, the wall must be cut to allow the faucet replacement. If you cannot cut

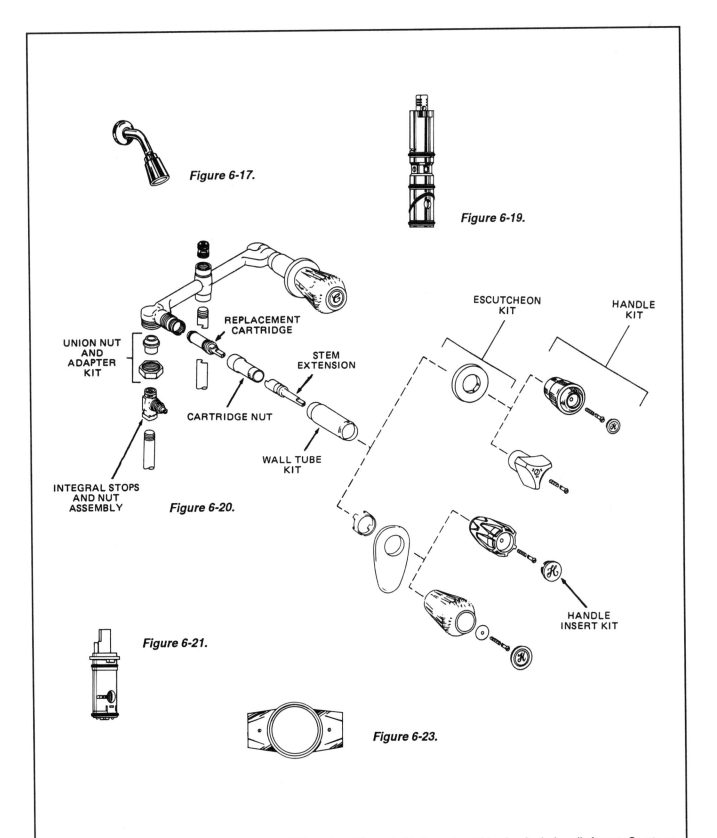

Figure 6-17.

Figure 6-19.

REPLACEMENT
CARTRIDGE

ESCUTCHEON
KIT

HANDLE
KIT

UNION NUT
AND
ADAPTER
KIT

STEM
EXTENSION

CARTRIDGE NUT

WALL TUBE
KIT

INTEGRAL STOPS
AND NUT
ASSEMBLY

Figure 6-20.

HANDLE
INSERT KIT

Figure 6-21.

Figure 6-23.

Figure 6-17. *Shower head and arm. Courtesy of Moen, Inc.* **Figure 6-19.** *Faucet cartridge for single-handle faucet. Courtesy of Moen, Inc.* **Figure 6-20.** *Multi-handle faucet parts. Courtesy of Moen, Inc.* **Figure 6-21.** *Multi-handle faucet cartridge. Courtesy of Moen, Inc.* **Figure 6-23.** *Remodeling cover plate. Courtesy of Moen, Inc.*

Figure 6-18. Tub wrenches.

Figure 6-22. Union.

the wall behind the faucet, you must cut the wall in front of it. The front wall is the wall of the tub or shower. Real problems can occur here.

If you cannot gain access from behind the faucet, call a professional. Plumbers will know how to gain access from the front of the faucet with minimal damage. If you make the wrong move in this endeavor, you could be looking at major repair costs for the opened wall. If you have a one-piece tub/shower combination, the hole can be devastating. With ceramic tile, the cost to repair it to match the existing tile can stagger you. Professional plumbers know what type of remodeling cover plates are available. Let the pros make the cut and take the responsibility.

Assuming you have access from behind the faucet, you will see two feed pipes coming into the faucet body. If you are working on a tub, you will see a pipe coming down from the body and turning 90° to penetrate the wall for the tub spout. If you have a shower head, you will see a pipe leaving the top center of the body, going up to the shower arm. All these pipes should be ⅝-inch pipe. This is commonly referred to as ½-inch pipe in the trade.

The pipe may be copper, galvanized steel, CPVC plastic, or polybutylene. Most homes will have copper pipe. This example is based on copper pipe since it is the most common.

There may be unions in the feed pipes or at the connection to the faucet body. Unions are couplings that may be unscrewed to separate the connection (Figure 6-22). If the pipes have unions, disconnect them by turning the nut counterclockwise. If you don't have unions, cut the feed pipes about 6 inches below the faucet body. Cut the pipe going up to the shower head about 6 inches above the body. Cut the pipe going to the tub spout about 3 inches above the 90° bend, between the bend and the faucet body. Disassemble the faucet trim from the front of the faucet. You can now remove the faucet body from the wall, through the access panel.

Try your replacement faucet body to be sure it fits the existing holes in the front wall. If all is well, continue with the replacement. Clean and flux the ends of the pipes you cut in the wall. Clean and flux four couplings. Install the couplings on the existing pipes. Cut sections of pipe to fit in the couplings and the faucet body. Clean, flux, and join the pipe to the couplings and the cleaned and fluxed fittings on the faucet body.

Some bodies come with threaded inlets. These require the use of male adapters. Put pipe compound on the threaded end of the male adapter and screw it into the inlet. Clean and flux the other end of the male adapter before inserting your pipe.

Open the valve or valves on the faucet body so the washers will not be burned during soldering. Check the alignment of the body and get ready to solder. Have a fire extinguisher at hand in case the walls catch on fire. You may want to spray water on the inner parts of the wall to reduce fire risk. When ready, solder the joints.

After the joints have cooled, close the faucet valves and cut the water on. Check all solder joints for leaks. If everything looks good, turn the faucets on and let them run. With water coming out of the tub spout, check the solder joint in the pipe feeding the spout. If it is okay, turn on the shower head and check the joint in its feed pipe. These two joints will not leak unless the water is running through them. Simply looking without turning the water on will not expose leaks in the feeds to the shower head or spout. If all is dry, you are done and can be proud of yourself.

If you have leaks, cut off the water at a point before the leaks. Drain the water from a lower point to clear the pipes of water. Apply flux around the leaking solder joint and solder it again. If you still have leaks, remove the bad joint by cutting it out and make a new joint with couplings.

Replacing Multi-Handle Faucets with Single-Handle Faucets

It is possible to replace two- and three-handle faucets with single-handle faucets. There are special remodeling cover plates available to cover the holes left by the old faucets (Figure 6-23). Before you attempt this, confirm availability of the cover plates. Also be sure to buy a faucet that is compatible with the cover plate.

Replacing a Tub Spout

There are two common types of tub spouts. The first type screws onto a threaded nipple or adapter. The second type is held in place on copper pipe with a set screw. The threaded type is the most common (Figures 6-24 and 6-25). To remove the threaded type, you just turn it counterclockwise. When you have it off, look to see where its threads are. They may be up inside the spout near the front or at the back. Get the same type of spout for your replacement. Apply pipe compound to the threads of the nipple or adapter and screw the spout on. It only needs to be installed hand-tight.

The style of spout with the set screw will have the screw under the spout, near the face of the tub. These are normally held on with hex-head screws. Loosen the screw and the spout will pull right off the pipe. Replace the spout with the same type and tighten the screw. Caulk around the spout where it meets the wall.

Tub Waste and Overflows

The tub waste is the piping running from the tub drain and the overflow hole to the tub trap. Tub wastes are normally made of plastic or brass. There are several different types of tub wastes; we will discuss three fairly common ones. They are a *trip lever waste* (Figure 6-26), a *twist-and-turn waste* (Figure 6-27), and a *push-down waste*.

The tub waste can be seen from below the tub if you can get below the tub and look up at it. The waste is generally accessible through a panel behind the tub. This is the same panel used to access the faucets. Again, there is not always an access panel for the waste. If the waste has solid connections instead of slip-nuts, access is not required by most codes.

Replacing Tub Waste and Overflow Washers

The tub waste may be held together with slip-nuts and washers. These slip-nut washers are replaced just as described for the kitchen sink. Loosen the nuts and separate the pipes. Remove the old washers and slide the new ones on. Reconnect the pipes and tighten the nuts.

Replacing the Face Plate Washer

Remove the face plate at the overflow hole in the tub by removing the screw or screws. There will be one or two large screws holding the plate in place. When the screws are removed, the plate will come off. Between the back of the tub and the surface of the overflow pipe is a gasket. Push on the overflow pipe

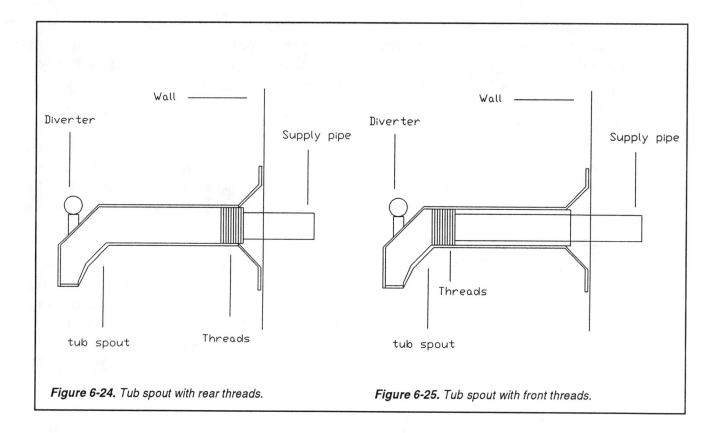

Figure 6-24. Tub spout with rear threads.

Figure 6-25. Tub spout with front threads.

and remove the overflow gasket. To install the new gasket, reverse the procedure.

Resealing the Tub Drain

The drain in your tub is sealed with putty on the tub side and a washer between the tub waste shoe and the tub. To repair these seals, the tub drain must be removed. Remove the strainer screen if there is one. It will be held in place by a single screw. Remove the screw-in stopper if there is one by turning it counterclockwise.

You can remove the tub drain with two strong screwdrivers. Put a screwdriver on either side of the cross bars and turn the drain counterclockwise. When the drain comes out, remove any existing putty from the drain. Remove the washer from between the shoe and the bottom of the tub. Install the new washer and put a ring of putty around the tub drain. Screw the drain back into the shoe until it is tight and the putty spreads out. This procedure stops water from leaking past the tub drain.

Replacing the Tub Waste and Overflow

If you have adequate access, this is not a bad job. Remove the tub drain and the overflow plate. Loosen the slip-nut where the tub shoe's drain enters the tee of the waste and overflow. Remove the tub shoe section. Loosen the slip-nut where the vertical drain enters the tub trap. If the overflow has a retainer ring at the overflow hole, remove it. Remove the remainder of the waste and overflow assembly. If the waste is glued together and made from plastic, you will have to cut it out instead of loosening slip-nuts.

To install the new waste and overflow, you will need pliers, screwdrivers, putty, and pipe compound. Apply pipe compound to the threads of the tailpiece. Screw the tailpiece into the bottom of the waste-and-overflow tee. Slide a slip-nut and washer on the tailpiece with the nut threads facing the trap adapter. Insert the tailpiece into the trap adapter and secure it with a slip-nut and washer. Take measurements and cut the overflow pipe to the proper length with tubing cutters or a hacksaw. Do the same for the shoe section.

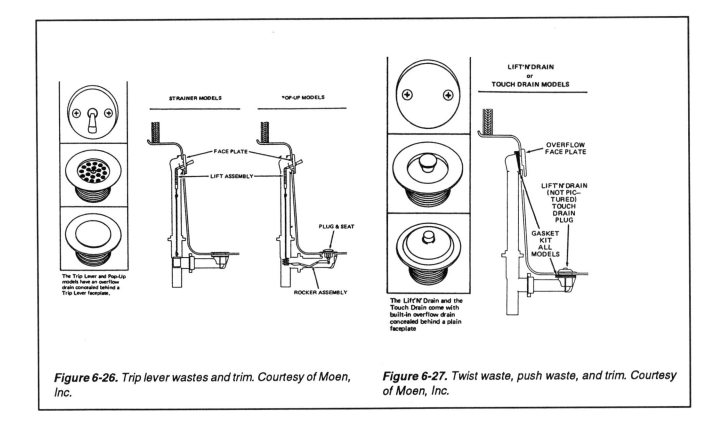

Figure 6-26. *Trip lever wastes and trim. Courtesy of Moen, Inc.*

Figure 6-27. *Twist waste, push waste, and trim. Courtesy of Moen, Inc.*

Slide a slip-nut and washer up on the overflow pipe with the threads facing the tee. Insert the overflow pipe into the tee and tighten the slip-nut. Follow the same steps for the shoe section. Now, go around and get in the tub with the rest of the parts you will need nearby. Install the shoe washer and the tub drain as discussed earlier. Install the overflow washer and face plate as discussed above.

The last step is to test your work. If you have leaks, tighten the slip-nuts and check the washers and gaskets. If you use a glue-together plastic waste, the only difference in installation is the glue joints instead of slip-nuts.

Adjusting the Tub Waste

If water is leaking out of your tub and down the drain, you need to adjust the tub waste. With the twist-and-turn and push-down styles, there is no adjustment procedure. These either work or they don't. If water leaks past these two types of stoppers, you must replace the stoppers.

Trip-lever wastes are adjustable. Remove the cover plate from the overflow and remove the overflow. The device attached to the trip-lever and face plate is what you adjust. The adjustment is normally made by turning the lower part of the waste on the threaded upper portion. There may be a nut on the threads that must be moved before you can turn the lower section on the threads. If water is leaking past your stopper, turn the lower portion counterclockwise. If the trip-lever doesn't go all the way down when depressed, allowing water to escape, turn the section clockwise. This adjustment is a trial-and-error adventure. You just keep working with it until you achieve the desired result.

7
Specialty Plumbing Fixtures

Some homes are equipped with special plumbing fixtures. These fixtures can include whirlpool tubs, spas, bidets, and other types of uncommon fixtures. This chapter provides a better understanding of these special fixtures. Some of these fixtures require the skills of a trained professional for repairs. Others have simple adjustments that the average person can learn to accomplish. Know your limitations and when in doubt, call a professional. Most of the fixtures covered here are expensive. If you don't feel confident, don't risk ruining the equipment to save the cost of a service call.

THE DIFFERENCE BETWEEN SPAS AND WHIRLPOOL TUBS

The major difference between a spa and a whirlpool tub is in the way they are used. Spas (Figure 7-1) are designed to hold the water in the unit for an indefinite period of time. Whirlpool tubs (Figure 7-2) are meant to be drained after each use. Spas do not connect to the sanitary plumbing drains of the home. They are drained with the use of a common garden hose. Spas are also filled with a garden hose; there is no need for formal plumbing to install a spa. A whirlpool tub is connected to the home's drainage system in the way a normal bathtub is. It is also connected to the potable water distribution system and filled with a faucet.

With a spa, chemicals are added to clean the water. Since whirlpools are drained after each use, they require no chemical additives. Another common difference is how the water is heated. A spa has an independent heater to heat the water. A whirlpool tub frequently depends on the home's water heater to heat the water. Some whirlpools are available with an optional heater, but they are less common than whirlpools without them.

If the unit is used regularly, heating water for a whirlpool can become expensive. The other drawback to heavy use of a normal whirlpool is the waste of water in draining the tub after each use. With a spa, the water remains in it for an extended period of time and is heated with the independent heater. This is more economical if the unit will be used often. Since a spa does not connect to the home's plumbing, it can be placed on a deck or in another location without sanitary plumbing.

Size is another difference between spas and whirlpool tubs. If you are looking for a large bathing unit, spas offer the most options. Whirlpools are most commonly designed to accommodate one or two people. There are larger versions available, but the water demands to fill them are heavy. Spas can easily accommodate up to four people or more. Since spas retain their water, they are less wasteful of water.

Gaining Access to the Whirlpool's Equipment

The equipment used to make a whirlpool work is normally hidden behind a false front. This false front is called an *apron*. The apron is a removable panel, mounted at the exposed side of the tub (Figure 7-3). This is usually where you step into the tub. The apron may be attached with screws or held in place with spring-tension clips. Refer to your owner's manual to determine the method of attachment for your apron. Most aprons require the bottom to be pulled out and the top to be pulled down. Even though this type is common, check your owner's instructions before attempting removal of the apron.

Gaining Access to the Spa's Equipment

Most spas are equipped with a full apron that surrounds the bathing unit. Some of these aprons are hinged and others are screwed into place. Check your owner's instructions for the recommended method for removing your spa's apron. Some companies refer to the apron as a *skirt*. If you see this wording in your instructions, it should have the same meaning as an apron. Normally, there is an illustration to show the proper method for accessing the unit's equipment.

The Parts of a Whirlpool Unit

The whirlpool configuration can appear confusing, but it is not overly difficult to understand. In the bathing unit is a *suction-intake fitting* (Figure 7-4). This fitting allows the water to be pulled into the system for circulation. Around the tub are several jets. These jets are where the water is dispersed into the tub by the whirlpool. The air-volume control is mounted near the top rim of the tub and is used to adjust the whirlpool action.

Under the tub, and behind the apron, is a maze of piping and other equipment (Figure 7-5). This is where the motor is located. The motor, pipes, and controls under the tub should only be worked on by trained professionals. The whirlpool equipment on spas is very similar to that of whirlpool tubs. In neither case should you attempt to work with the equipment behind the apron.

The Drainage Connection of a Whirlpool Tub

Whirlpool tubs are connected to the sanitary plumbing system of the home. This is done with a tub waste and overflow. The drain size is $1\frac{1}{2}$ inches in diameter. This is the same type of tub waste found on regular bathtubs. The only difference may be in the height of the overflow tube. Since whirlpool tubs are often deeper than standard tubs, many of them require a taller overflow tube. The adjustment and replacement of these tub wastes and overflows are the same as the instructions given in Chapter 6 for standard bathtubs.

Whirlpool Faucets

The faucet on a whirlpool is a little different from a regular tub faucet. The faucet may be built into the tub, deck-mounted on the tub, or wall-mounted. The spout is long and usually a part of a three-part faucet set. This type of faucet is usually called a *tub filler*. In many cases, these tub fillers are equipped with a personal shower unit. Figure 7-6 shows a typical deck-mounted tub filler and the associated parts. The repair and replacement of these faucets is similar to the instructions given in Chapters 5 and 6 for faucets. Since there are countless variations in this type of faucet, refer to the owner's manual for detailed repair and replacement instructions.

There are a few tips you should know about three-piece faucets. The small tubing that connects the handle valves to the tub filler can be difficult to work with. If the tubing is copper, it will crimp very easily. Use caution when bending this type of tubing. The tubing can also become damaged if you apply too much stress to it by turning the handle valves during installation. Be sure the tubing is not being bent as you tighten the handle valves. This tubing is normally connected to the handle valve and the tub filler with compression fittings or a soldered joint. If the tubing is soldered into the fitting, be careful not to break the solder joint. This damage can occur by the same means, causing the tubing to crimp.

When a personal shower attachment is used, a vacuum breaker should be installed with the tub-filler assembly. Vacuum breakers prevent the water

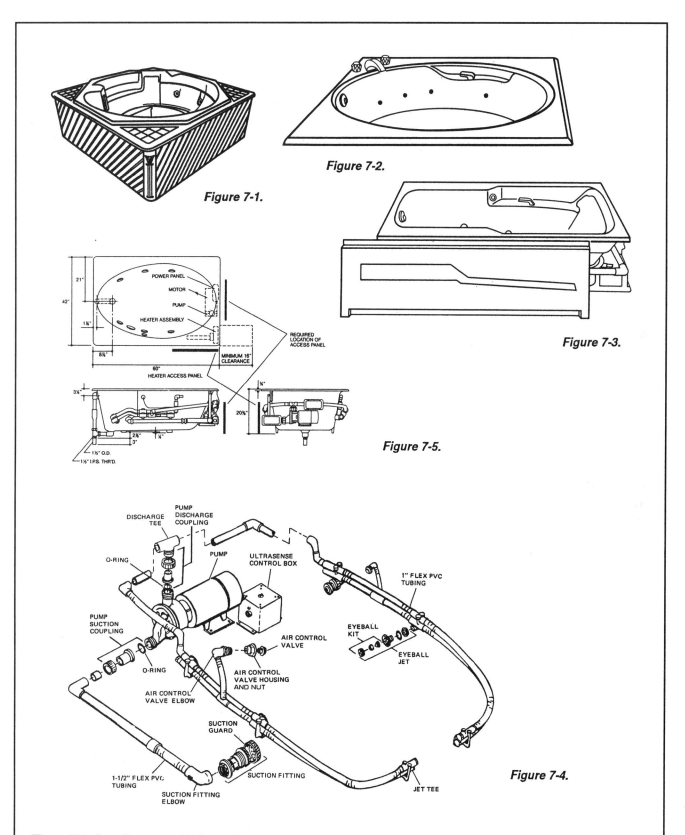

Figure 7-1. Spa. Courtesy of Universal-Rundle. **Figure 7-2.** Whirlpool tub. Courtesy of CR/PL, Inc. **Figure 7-3.** Apron. Courtesy of CR/PL, Inc. **Figure 7-4.** Parts of a whirlpool. Courtesy of American Standard, Inc. **Figure 7-5.** Cross-section of whirlpool equipment. Courtesy of Kohler Company.

in the bathing unit from being siphoned back into the potable water supply. A vacuum breaker is a simple device that is usually required by the plumbing code (Figure 7-7). The vacuum breaker receives the water in the inlet and releases it through the outlet. If a pipe in the main plumbing system were to break, it could cause a vacuum or siphon effect. Vacuum breakers prevent the water in the tub from being sucked through the personal shower and back into the potable water pipes. Since personal shower units are sometimes left in a tub filled with water, the vacuum breaker is a good safety device.

Whirlpool Timers

If your whirlpool is equipped with a timer, call a professional when the timer requires work. There is nothing in the timer device for the average home-owner to work on.

Whirlpool Tub Replacement

Replacing a built-in whirlpool tub is quite a job. It involves many aspects of work and is too complicated for the average homeowner to attempt. The job not only involves plumbing, but also electrical wiring and wall repair, and it can stretch into floor work.

Spa Replacement

Replacing a free-standing spa is heavy work, but simple. Disconnect the electrical wire and drain the existing spa. The draining is usually done with a garden hose attached to the boiler drain found inside the apron. When the spa is empty, you can move it. Expect to need help when moving the spa. Depending on the size of the unit, two to four people will be needed to move the spa.

Once the old spa is out of the way, place the new spa where you want it. Follow the manufacturer's suggestions for setting up the spa. Normally, all that is required is filling the spa with a garden hose and plugging in the electrical cord. If the spa is to have special accessories, such as a timer, call a professional to handle the complex aspects of the installation.

SOAKING TUBS

Soaking tubs are oversized bathtubs without whirlpool jets (Figure 7-8). These tubs are designed to allow the occupant to soak in deep water without the whirlpool action. This type of tub can be recessed into the floor or set on the floor with an apron surrounding the tub. There are no motors, controls, or whirlpool piping to be concerned with.

Soaking tubs are connected to the plumbing with the same principles as a regular bathtub. The common difference is the height of the tub waste and overflow and the style of faucet used. The waste and overflow are higher than normal, like the ones discussed in the whirlpool section. The faucets are the tub-filler type, also discussed in the whirlpool section. Basically, soaking tubs are plumbed as a whirlpool is, except for the whirlpool jets and related equipment.

Replacing a Soaking Tub

A soaking tub is a big job to replace. Unless you are ready to get into serious work, call in a professional for the replacement of a soaking tub. The replacement of the tub will result in extensive work of various natures.

BIDETS

Bidets are not a common plumbing fixture in most American homes, but they are found in some. The bidet is a personal hygiene fixture (Figure 7-9). For such a simple-looking fixture, bidets can become challenging very quickly.

A bidet sits on the floor and has a drain and a faucet. The faucets used with bidets can have a staggering number of parts. The drain works on the same principle as a lavatory drain. The bidet is connected to the sanitary plumbing by way of a slip-nut and washer. The drain is $1\frac{1}{4}$ inches in diameter.

At a casual glance, bidets do not seem intimidating, but they can be complicated. The faucets can be deck-mounted or wall-mounted. The drain assembly can be confusing, and the fixture can frustrate the untrained homeowner very quickly. There is usually a vacuum breaker to add to the complications. Many bidets are equipped with a spray unit in the bowl. Before doing any work on your bidet, refer to the manufacturer's manual.

There are numerous connections and fittings used to make the bidet operational (Figure 7-10). The drain has a tailpiece, and the connections can be uncomfortable to access. With all these obstacles, you may want to consider calling a trained technician to repair or replace your bidet.

Common Problems

On the whole, there is little need for maintenance with a bidet. The drains hardly ever become clogged, and there are no motors or controls to fuss with. The pop-up assembly is like that of a lavatory. It is unusual that these assemblies need adjusting. The faucets may need occasional attention to change a washer or replace a cartridge, but they are worked on like normal faucets. The vacuum breaker should not require work, but if it does, replacement is not difficult. The spray in the bowl should not demand any effort once it is adjusted during the initial installation. The slip-nut washer on the drain may have to be replaced after years of service, but again, this is similar to working with a lavatory.

Bidet Faucets

The faucets for a bidet can be wall-mounted or deck-mounted. They can have single or multiple handles. The repair and replacement procedures for these faucets are the same as for the faucets already described throughout this book. There are two things you may find different about a bidet faucet. One is the vacuum breaker, and the other is the spray connection.

Over-rim bidet faucets. Some bidets use an over-rim faucet. These faucets do not utilize a vacuum breaker because they do not have a spray unit in the bowl. This type of faucet is the exception rather than the rule with bidets.

The Vacuum Breaker

The vacuum breaker extends up above the faucet. The vacuum breaker generally originates at the center of the faucet body. There is a short nipple screwed into the valve body. At the other end of the nipple is an elbow. From the elbow, there is another short nipple to which the vacuum breaker is attached. Some vacuum breakers are attached to a flexible hose instead of a rigid nipple.

Replacing the vacuum breaker. Refer to your owner's manual to see how your vacuum breaker is attached and what type it is. The replacement of vacuum breakers is normally a simple procedure. In average conditions, all you have to do is unscrew the old vacuum breaker and replace it with a new one. Pipe compound is needed on all threaded fittings when making the replacement. With some bidets, the vacuum breaker is mounted to the wall. The replacement procedure is the same.

If your bidet has a large, ornamental chrome tube extending upward, it houses the vacuum breaker. Refer to the manufacturer's instructions for gaining access to the vacuum breaker. This is typically accomplished by unscrewing the chrome housing.

The Spray Connection

The spray unit in the bowl of the bidet is connected to the faucet. For simplicity, you can think of the spray unit as being the tub spout on a tub-faucet setup. There is no need for routine maintenance to the spray unit. If it needs work, refer to your owner's manual. The spray unit is attached to a transfer valve that diverts the water to the spray. You can think of this as the diverter that sends water to the shower head instead of the tub spout. Plumbing is a matter of principles. Once you understand the basics, you can usually work with any normal plumbing situation.

Bidet Pop-Up Drains

The pop-up drain on a bidet is very much like the pop-up drain for a lavatory. Some bidets have a curved lift rod, but the mechanism is the same.

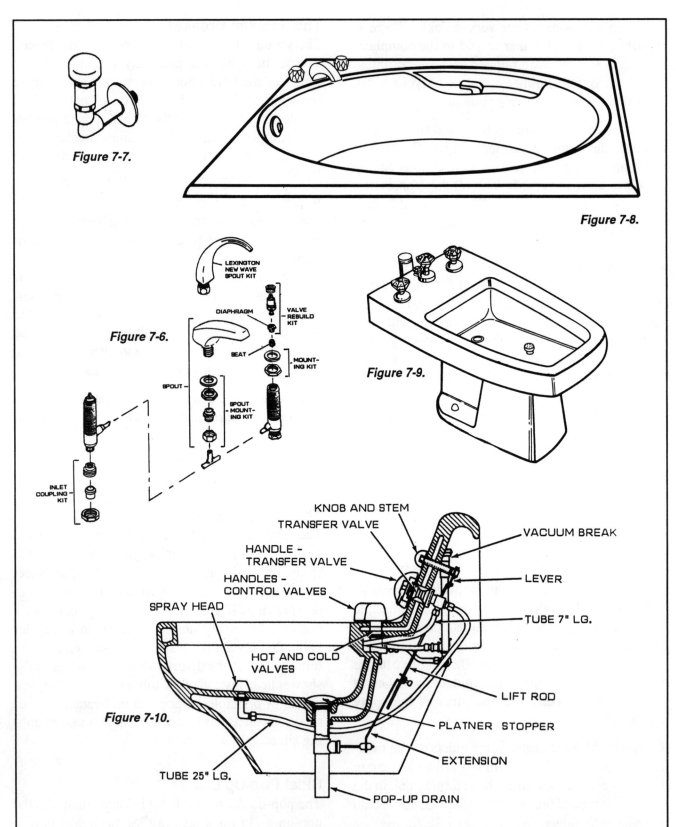

Figure 7-6. *Deck-mounted tub filler and parts. Courtesy of American Standard, Inc.* **Figure 7-7.** *Tub-filler vacuum breaker. Courtesy of American Standard, Inc.* **Figure 7-8.** *Soaking tub. Courtesy of CR/PL, Inc.* **Figure 7-9.** *Bidet. Courtesy of Universal-Rundle.* **Figure 7-10.** *Cross-section of bidet. Courtesy of American Standard, Inc.*

Repair and replacement procedures are the same as those discussed in Chapter 6 for lavatories. The only difference is the access point to the pop-up. With most bidets, the pop-up is reached from behind the fixture.

Replacing a Bidet

If you are good at following instructions, replacing a bidet is not difficult. Find the fixture's shut-off valve and cut it off. Look inside the bidet, from behind, and locate the slip-nut connection. Run out any existing water in the pipes and prepare for the replacement.

Loosen the slip-nut connection to free the bidet's drain. Remove the covers from the mounting bolts on the bidet. These are similar to those of a toilet. They are located on the outside flange of the bidet, near the center of the drain. Loosen the bolts. Disconnect the water connections from the faucets. With deck-mounted faucets, remove the faucets. When all connections are loose, the bidet should lift up and be easy to remove.

Set the new bidet in place and reverse the removal procedure. If the replacement bidet is compatible, the new installation is simple. If the water connections don't line up, you will have to adjust for the differences. This can be done with couplings and other fittings. The drain connection is based on the same principle as that of a lavatory.

Install the pop-up assembly and connect the tailpiece to the drain with a slip-nut. This connection is made from behind the bidet. Replacing a bidet is not a major chore. The drainage connection should be simple, and the water connections are easy to maneuver. Obviously, it is easiest to replace the existing bidet with one of similar design. When the replacement is complete, cut the water valve on and check for leaks. Run the fixture and check the drain for leaks. By following the manufacturer's suggested installation procedures, you should not encounter any insurmountable obstacles.

CLAW-FOOT BATHTUBS

Claw-foot bathtubs are not like regular tubs. They are not surrounded on three sides by walls, and their plumbing is different. Faucets for a claw-foot tub mount directly into the tub. The tub waste is a different size from normal tub wastes, and the overflow tube is a different size. When you are working with an old claw-foot tub, the fittings can be difficult to loosen and work with.

Faucets for a Claw-Foot Tub

The faucets used on old claw-foot tubs have two handles and a small spout that sticks straight down toward the tub. These faucets are mounted to the tub and sit below the flood-level rim of the tub. By today's plumbing code, these faucets are not legal in most jurisdictions. The fact that the spout is below the flood-level rim makes it illegal.

Replacing the washers in these faucets is easy if the faucet will come apart. Many of these old faucets become seized and are uncooperative. Cut off the water and remove the faucet's handles. Use a wrench to loosen the retainer nuts around the faucet stems. When the retainer nut is loose, the stem will come out. Once the stem is out, you can change the washer, replace the faucet seats, or replace the packing around the stem. This type of faucet is similar to the standard two-handle faucets covered in Chapter 6.

Removing the Faucet from an Old Claw-Foot Tub

Cut off the water and loosen the unions at the faucet inlet connections. Loosen the mounting nuts that are around the faucet inlets and close to the back of the tub. The mounting nuts unscrew down the length of the inlet connection. When these nuts are off, the faucet will pull out of the front of the tub.

Replacing the Faucet for a Claw-Foot Tub

Since the old claw-foot faucets do not comply with most plumbing codes, you should not use this type of faucet as a replacement. The proper replacement faucet for a claw-foot tub is called a *code faucet* within the trade. This faucet mounts in the same holes as the old faucet, but the spout is made in a gooseneck fashion.

This design gets the spout of the faucet above the flood-level rim of the tub. The faucets are placed through the holes in the claw-foot tub. Then the mounting nuts are threaded onto the inlet connections of the faucet. The next step is attaching the water supply pipes to the faucet. Special elbows are sold for connecting the water supply pipes to the faucet. These elbows have female threads on the bottom and a union-type nut where the connection is made to the faucet. If you tell your plumbing supplier that you are connecting the water supply to a claw-foot tub, the supplier will know which parts to give you. If age and rust have not sealed the old faucets to the tub, replacing claw-foot faucets is not difficult.

Waste and Overflow Assemblies for Claw-Foot Tubs

If you have the occasion to work with the waste and overflow of a claw-foot tub, you will find it is an odd size. A standard tub waste and overflow is too large to work on a claw-foot tub. The tubing on the waste and overflow for a claw-foot tub has a diameter of $1^3/_8$ inches. A standard waste and overflow has a diameter of $1^1/_2$ inches.

These odd-sized waste and overflows are still available. You may have to do some looking around, but you can find them. The mechanics of repairing or replacing the waste and overflow is the same as on a standard waste and overflow. The key difference is in the overflow tube. A claw-foot tub waste and overflow uses a stopper-plug to hold water in the tub. It does not depend on a trip lever or a toe-touch stopper. It works with a simple stopper that is placed in the tub drain. The face-plate of the overflow tube is usually just a wire-mesh cover.

PEDESTAL LAVATORY

The key difference between a pedestal lavatory and a standard lavatory is the method of support for the lavatory bowl (Figure 7-11). The workings of the waste and water lines are the same. The faucets can be 4-inch center faucets or 8-inch center faucets, but they are worked on the same way as the others

discussed throughout this book. There is no real difference in repair procedures, but there are differences in the replacement techniques.

Pedestal sinks hang on a wall bracket, much like a wall-hung lavatory. The bracket supports most of the bowl's weight. The pedestal is a separate part. It adds support and hides the pipes attached to the bowl. A few manufacturers design their pedestals to be bolted to the floor. The bowl may be bolted to the wall. Before attempting the removal of these expensive lavatories, do some research.

When possible, follow the instructions in your owner's manual for working with pedestal sinks. Look under the bowl to see if it is bolted to the wall. Some bowls simply sit on the wall bracket; others sit on the wall bracket and are bolted to the wall.

Removing a Pedestal Lavatory

Assuming the bowl is mounted to the wall, the pedestal should be able to be removed without the bowl falling. I cannot guarantee the bowl will not fall, but it shouldn't. Obtain an assistant to help you with the removal. Your assistant can support the bowl while you remove the pedestal. This way, if the bowl tries to fall, your assistant can catch it.

Look inside the pedestal to see if it is bolted to the floor. Most pedestals are held in place by the weight of the lavatory bowl and are not anchored to the floor. If there is a lag bolt there, remove it. Have your assistant apply slight upward pressure on the bowl. While the bowl is held up, remove the pedestal. It should just slide out from under the bowl.

With the pedestal out of the way, you have clear access to the waste and water connections. Disconnect these pipes by following standard procedures. When the waste and water lines are loose, lift the bowl off the wall bracket.

Installing the New Pedestal Lavatory

Installing the new pedestal lavatory is a matter of reversing the removal procedure. Mount the new wall bracket. Install the faucets and pop-up assembly on the new sink. Hang the sink on the wall bracket and connect the waste and water connec-

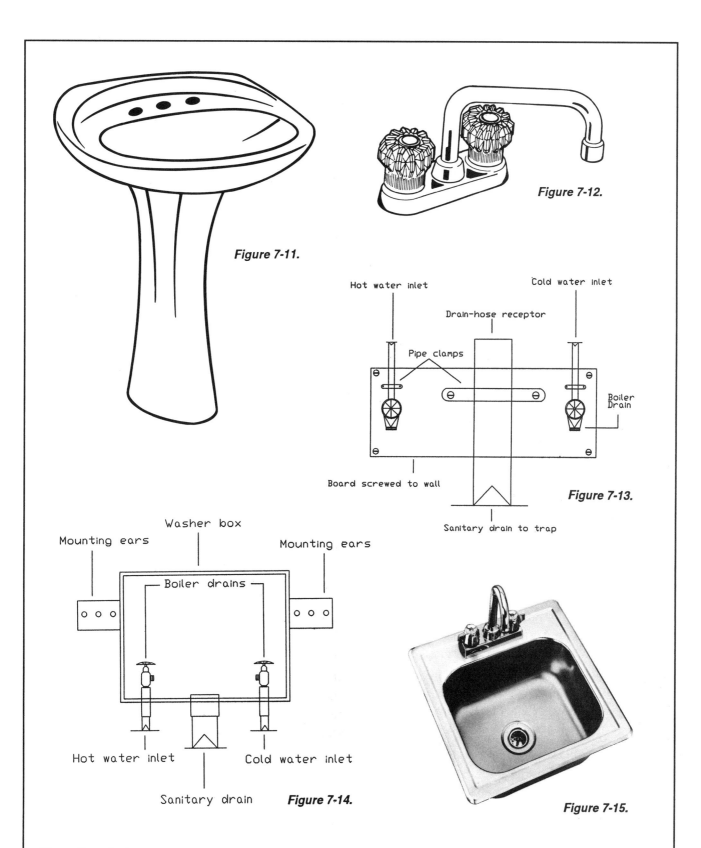

Figure 7-11.

Figure 7-12.

Hot water inlet

Cold water inlet

Drain-hose receptor

Pipe clamps

Boiler Drain

Board screwed to wall

Sanitary drain to trap

Figure 7-13.

Washer box

Mounting ears

Mounting ears

Boiler drains

Hot water inlet

Cold water inlet

Sanitary drain

Figure 7-14.

Figure 7-15.

Figure 7-11. *Pedestal lavatory. Courtesy of CR/PL, Inc.* **Figure 7-12.** *Deck-mounted laundry faucet. Courtesy of Universal-Rundle.* **Figure 7-13.** *Boiler-drain wall hookup for washing machine.* **Figure 7-14.** *Washer outlet box.* **Figure 7-15.** *Bar sink. Courtesy of UNR Home Products.*

tions. Place the pedestal under the bowl and you are done.

The installation is not always as simple as it sounds. Most pedestals depend on the weight of the bowl to hold them in place. This means the wall bracket must be mounted at the precise height to maximize the effectiveness of the design. Follow the installation instructions provided with the new lavatory to make sure all works well.

LAUNDRY TUBS

Laundry tubs are not a exotic fixtures, but since they are not found in all homes, I refer to them as specialty fixtures. The laundry tub is sometimes called a laundry tray or a utility sink. These sinks are deep and are normally installed near a washing machine. A normal laundry sink is about 2 feet wide and 1 foot deep.

Laundry sinks are used to soak stained laundry, to accept the waste water from a washing machine, to bathe pets, for photo darkrooms, and in countless other applications. The sinks and their faucets are simple to work with. There are two common types of laundry tubs. One hangs on the wall, and the other is supported by legs. Modern laundry tubs are usually made of plastic or fiberglass.

Deck-Mounted Laundry-Tub Faucets

Deck-mounted faucets are very common on modern laundry tubs (Figure 7-12). The faucet fits through holes on a 4-inch center. The faucets are secured with typical faucet-retainer nuts. The connections under the sink are the same as on a standard lavatory or kitchen faucet. There is no pop-up assembly with laundry faucets. Laundry tubs use a stopper-plug to retain water in the sink.

The areas for maintenance in a laundry faucet are about the same as those of a kitchen faucet. There are generally faucet washers, packing, and faucet seats that may be repaired or replaced. The procedure for this work is the same as that of any standard faucet. The faucet's spout is held in place by a knurled nut. There is an "O" ring around the spout where it fits into the faucet body. This "O" ring may need to be replaced from time to time.

Replacing the "O" ring is easy. With the faucets turned off, loosen the knurled nut around the spout. When the nut is loose, lift the spout out of the faucet. Peel off the old "O" ring and install a new one. Set the spout back into the faucet and tighten the retainer nut. This will prevent water from leaking out around the spout. If the entire faucet needs to be replaced, follow the instructions given in Chapters 5 and 6.

Lip-Mounted Laundry Faucets

Some laundry sinks, especially old ones, are equipped with a faucet that mounts on the lip of the laundry sink. These faucets receive a $1/2$-inch water supply from above the sink. Unlike deck-mounted faucets, these faucets are connected directly to the water pipes, instead of to $3/8$-inch supply tubes.

Instead of using conventional mounting procedures, these faucets are mounted to the backsplash rim of the laundry tub. The faucets are held in place by set screws. The internal mechanisms are about the same, but replacement procedures are different.

Before you can replace these faucets, you must cut off the incoming water. This may need to be done at the home's main shut-off valve. When the water is cut off, open the faucets to drain any existing water. Loosen the unions in the water pipes, if you have them. If you do not have unions, cut the pipes. Loosen the set screws holding the faucet to the sink. Remove the faucet.

Set the new faucet in place and secure the set screws. This type of faucet requires the use of soldered joints. When the faucet is set, open the faucet valves. If the valves are closed during the soldering process, the rubber washers may melt. Install your new pipe in the faucet and connect it to the existing water pipes. When everything is connected, cut the water back on and check your connections for leaks.

Drainage Leaks

Repairing drainage leaks with a laundry tub is done with the methods described in Chapters 5 and 6.

Replacing a Wall-Hung Laundry Sink

Wall-hung laundry sinks are just overgrown wall-hung lavatories. The replacement process is very similar. The sink hangs on a wall bracket, secured to the wall. With the waste and water connections disconnected, the sink should lift off the wall bracket. Replacing the sink is as simple as reversing the removal procedure. Refer to the instructions given for wall-hung lavatories in Chapter 6.

The only big difference is the installation of the sink's drain. Unlike a lavatory, laundry sinks do not have pop-up assemblies. The drain is simply mounted to the sink and blocked with a rubber stopper. Before installing the drain, place plumber's putty around the rim of the drain where it will come into contact with the sink. Push the drain into place until the putty spreads out.

From below the sink, slide the fiber washer over the threaded portion of the drain. Follow the fiber washer with the mounting nut. Tighten the mounting nut, and you are ready for the tailpiece. The flanged tailpiece is held in place with a slip-nut and washer. From here on, the installation is the same as for any standard sink.

Replacing Free-Standing Laundry Sinks

Free-standing sinks are supported by four legs. They are not mounted to a wall bracket. After the water is cut off, disconnect the waste and water connections. The sink lifts out of the way to make room for the replacement sink.

Installing the new sink is simple. Push the four legs into the receiver holes on the bottom of the sink. When the legs are in place, set the sink where you want it. The tips of the legs are adjustable. By turning the leg-tips, you can level the sink. When the sink is level, install the faucet. Make all the plumbing connections in the normal methods, and you are ready to test your work.

WASHING MACHINE HOOKUPS

Most homes provide for the connection of a washing machine, but the types of connections can take many forms. They can be as simple as a drain pipe and two $1/2$-inch boiler drains, attached near the wall (Figure 7-13). The connection might utilize a box to house the waste and water connection (Figure 7-14). Another type of connection uses a faucet as a hook-up for the washing machine hoses. Since this type of plumbing is not a code requirement, it might be considered a specialty fixture.

Whether you have a washer box or have your connections attached near the wall, repair procedures are the same. Washing machines use a hose to connect to the water supply. The connection of these hoses is like that of a garden hose. The valve these hoses connect to is called a *boiler drain*.

Boiler drains have external threads for the hose to connect to. The connection is made by a fitting that screws onto the threads. The fitting is a part of the hose and has the same thread pattern as a garden hose. Inside the ring that screws onto the boiler drain is a washer. The washer is flat and is the same type as is used with a garden hose.

Replacing the Washers in a Washing Machine Hose

When these washers need replacing, the job is simple. Cut off the boiler drain and unscrew the hose's knurled nut. When you look into the knurled nut, you should see a round washer. Remove the washer and push a new one in to replace it. Garden-hose washers are not held in place by screws like faucet washers. They just sit inside the nut and against the hose.

Replacing Boiler Drains

Boiler drains come in different styles. Some of them screw into female threads, and some are equipped with compression fittings. The ones that screw into female threads are usually designed to accept a solder joint as well. The replacement of the boiler drains depends on the type in use.

Look at the existing boiler drain and determine its method of attachment. If the existing valve is screwed into female threads, plan on screwing in the replacement. First, cut the water off to the pipes supplying water to the boiler drains. Disconnect the washing machine hoses by turning the connecting nut counterclockwise.

Use a wrench to turn the boiler drain counterclockwise. It may be necessary to put a second wrench on the fitting into which the boiler drain is screwed. This will keep the solder joints from breaking as you turn the boiler drain. When the old valve is out, you are ready to install the replacement.

Apply pipe compound to the threads of the replacement boiler drain. Screw the new valve into the existing location by turning it clockwise. When the boiler drain is hand-tight, tighten it with your wrench. Again, it is a good idea to have a second wrench on the fitting into which the boiler drain is being screwed.

When the new boiler drain is tight and properly aligned, attach the washing machine hose. Cut on the water to the boiler drain and open the valve of the boiler drain by turning the handle counterclockwise. Start your washing machine and inspect the installation for leaks. If the fitting leaks, try turning the boiler drain a little tighter.

Replacing Compression-Type Boiler Drains

Cut off the water supply to the boiler drains. Disconnect the washing machine hoses from the boiler drains. Grasp the throat of the boiler drain with a pair of pliers. Put your wrench on the large compression nut, at the base of the boiler drain, and turn it counterclockwise. The pliers will prevent the boiler drain from turning to allow the loosening of the compression nut. When the compression nut is loose, remove the old boiler drain.

To install the new compression-type boiler drain, reverse the removal procedure. Hold the new boiler drain in place and hand-tighten the compression nut. When the nut is hand-tight, hold the boiler drain with pliers and tighten the compression nut with a wrench. When finished with the replacement, cut the water supply back on to test for leaks.

Replacing Solder-Type Boiler Drains

When your boiler drains are soldered onto copper pipe, you have two options. You can cut the pipe and install the new boiler drains using couplings and small pieces of pipe. The other option is removing the existing drains by sweating them off. This method requires less work and fewer fittings.

To sweat the old valves off, you will need your soldering gear and a fire extinguisher. Cut off the water supply to the boiler drains and disconnect the washing machine hoses. If there is some point in the water distribution system with an outlet below the boiler drains, your job will be a little easier. When possible, shut off all the water to the home and open a faucet located below the boiler drains. This will allow the water standing in the pipes to drain out.

When the water supply for the boiler drains comes in from above, opening the boiler drain will empty the pipe. Even if you are unable to drain the water from the pipe, you may be able to steam it out. With the water off, open the valve on the boiler drain and apply heat from your torch to the boiler drain's soldered joint. Don't stand in front of the boiler drain. If water is present, scalding steam may shoot out the end of the boiler drain.

When you see the solder turn soft, use a pair of pliers to remove the boiler drain. You may have to twist or turn the boiler drain to get it off the pipe. When the boiler drain is off, apply heat to the solder remaining on the pipe. When it is fluid, use a thick rag to wipe off the excess solder. Be careful, as the pipe and fittings will be very hot.

Now you are ready to install the new solder-type boiler drains. Use sandpaper to sand the pipe before installing the boiler drain. Apply a liberal coating of flux to the pipe. Open the valves on the new boiler drains, so the washers will not be burned during soldering. Clean the inside of the boiler drain's fitting socket and apply flux to the cleaned surface. Attempt to place the new boiler drain on the existing pipe. You may find it will not slide onto the pipe. If this happens, you will have to heat the pipe to melt more of the old solder.

When the old solder is fluid, place the new boiler drain on the pipe using pliers. When the new drain is seated on the pipe, solder the connection. When you are done and the pipe and fittings have cooled, close the valve on the boiler drains. Close any other faucets you may have opened and turn the water back on to check for leaks.

Washing Machine Faucets

The faucets used for washing machines are very inexpensive. If your laundry faucet is giving you problems, don't waste time trying to repair it. Replace the faucet with a new one.

These faucets are made to accept a standard $5/8$-inch pipe in their pipe sockets. This pipe is know in the trade as $1/2$-inch pipe. Follow the directions given above for replacing solder-type boiler drains. The only difference in the replacement is that the faucet has the hot and cold all in one body. The technical procedures are the same as for soldered boiler drains.

Replacing the Trap at Your Washing Machine Connection

Most washing machine hookups have a 2-inch trap. In some states, it is legal for the trap to be $1^{1}/_{2}$ inches in diameter, but most are of the 2-inch variety. Replacing the trap is a simple job, only requiring you to cut the pipe and replace the trap.

Normally, the washing machine trap is a "P" trap. This is a receptor-type connection, meaning the washing machine's discharge hose is not directly connected to the trap. It merely sits inside the receptor pipe. The trap may be connected to the drainage system with a slip-nut and washer, but most are connected with regular drainage fittings. You will generally have to cut the pipe at the trap arm to replace the trap.

You can use a coupling to connect the new trap arm to the existing drainage. If you do have a slip-nut and washer, all you have to do is loosen the slip-nut and remove the trap arm. The receptor pipe should be at least 18 inches high, but it should not be more than 30 inches high.

BAR SINKS

Most bar sinks are just miniature kitchen sinks (Figure 7-15). If you refer to Chapter 5, you can use the information given for kitchen sinks to work with your bar sink. The sink mounting, faucet mounting, and drain connections are all done in the same manner. The only noticeable difference other than size is the gooseneck faucet on the bar sink. Even though the faucet has a gooseneck spout, the repair and replacement procedures are the same as with the traditional faucet.

8
Water Heaters and Domestic Heating Coils

Hot water is an expected comfort in our society. Many people take their hot water for granted. They assume that when they turn on the faucet, there will be hot water available. When there is no hot water, they are perplexed and angered. The average home-owner has no idea how to determine why he or she doesn't have hot water. This chapter is dedicated solely to hot water and the devices that provide it.

Your hot water can be generated by a number of devices. Depending on the equipment in your home, hot water may come from:

• Electric water heater
• Gas-fired water heater
• Domestic heating coil
• Oil-fired water heater

With so many possibilities for hot water, you must identify your source before troubleshooting it. There are similarities to the various water heaters, but there are significant differences too. As you go through this chapter, it is broken down into sections for each type of water heater. The first heater described is an electric water heater.

ELECTRIC WATER HEATERS

The first thing you should know about electric water heaters is their ability to electrocute you. Most residential water heaters receive 220 volts of electricity. This is a lot of voltage, and it can be fatal if encountered. *Never, I repeat, never work on an electric water heater without being acutely aware of the electrical current.*

Leaking Relief Valves

The relief valve is an important part of your hot water heater. It is a safety valve that protects against excessive temperature or pressure within the water tank (Figure 8-2). If you removed the relief valve and plugged the hole, you could be sitting on a bomb. Water heaters can and have exploded. The force from these explosions is dramatic and can cause serious personal and property damage. Never remove a relief valve and replace it with a plug. Always replace the relief valve with a new valve of the same rating. You will find the rating stamped on a metal tag around the handle of the relief valve.

To remove the existing relief valve, cut off the electrical power to the water heater. Turn off the water that is feeding into the water heater. There should be a valve on the cold water pipe, near the water heater. Drain water from the tank until the water level is below the relief valve. To drain the tank, turn the drain handle counterclockwise. To speed up the draining process, open the handle on the relief valve by pulling it up. This allows air into the tank so it will drain faster.

There should be a pipe running from the relief valve to within about 6 inches of the floor. Remove this pipe by turning it counterclockwise. When the water level is below the relief valve, turn the valve counterclockwise. A pipe wrench is the best tool for this job. Apply pipe compound to the new relief valve and screw it into the tank where you removed the old one. Tighten the valve with the pipe wrench and reinstall the discharge pipe you removed from the old valve.

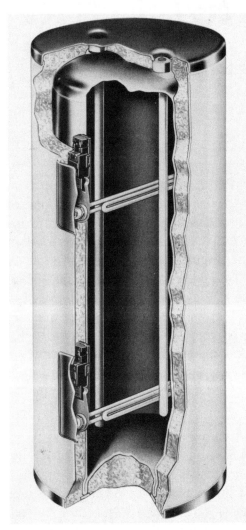

Figure 8-1.

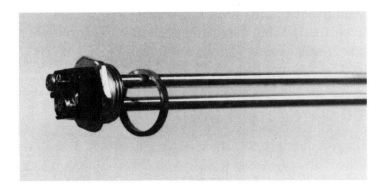

Figure 8-2.

Figure 8-3.

Figure 8-1. *Electric water heater in cutaway view. Courtesy of A.O. Smith.* **Figure 8-2.** *Relief valve.* **Figure 8-3.** *Screw-in element.*

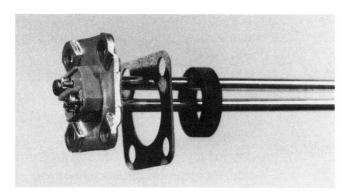

Figure 8-4.

Figure 8-5.

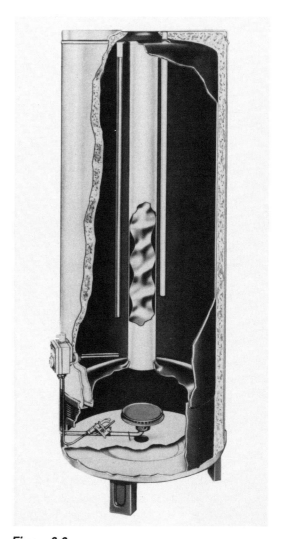

Figure 8-6.

Figure 8-4. Bolt-in element. Figure 8-5. Thermostat. Figure 8-6. Gas-fired water heater in cutaway view. Courtesy of A.O. Smith.

With the drain of the water heater closed, turn the water back on to the heater. Pull up the handle on the relief valve to allow air to escape. When water comes out of the relief valve, close the handle and turn the electricity back on. *Never turn the electricity on when the water level in the tank is low.* If there is no water covering the heating elements, they will burn out and need to be replaced.

Drains that Won't Drain

Water heaters build up a sediment layer at the bottom of the tank. This sediment can block the drain, preventing the heater from draining when the drain is opened. If this happens, cut off the power to the heater. Open the drain and probe through the drain hole with a stiff piece of wire, such as a coat hanger. If it still will not drain, you will need your pipe wrench.

This next step will cause the water in the tank to flood out from the drain hole. There is no way to control the water once this procedure is done. If you don't want to flood the area, don't attempt this task. With the drain all the way open, you should see a steel nipple going into the tank. The nipple has the drain handle on one end and screws into the tank at the other end.

Put the pipe wrench on the nipple and turn it counterclockwise. When the nipple is loose, water should rush out. It is rare that the sediment is heavy enough to block the nipple hole completely. This is the hard way to drain a tank, but sometimes it is the only way. Before reinstalling the nipple, apply pipe compound to the threads.

Access Panels

Behind the access panels of a water heater are elements, thermostats, and temperature control settings. There is usually 220 volts of power behind these panels. The wires are often hidden by fiberglass insulation, but they are there and they can be deadly. It is unwise for anyone other than a trained professional to remove these access panels.

Heating Elements

The heating elements in electric water heaters come in two different styles and with different wattage ratings. Most electric water heaters have two elements of one type or the other: *screw-in elements* (Figure 8-3) or *bolt-in elements* (Figure 8-4). These elements extend into the tank and use electricity to heat the water.

When you have hot water but it doesn't last very long, you probably have a bad lower heating element. If you have no hot water, the upper element may be burned out. Checking and replacing heating elements is a job for a professional. Working with the elements requires direct contact with high voltage and can be very dangerous. I cannot advise you on how to do this job, due to the extreme risks of personal injury. I will, however, give you an idea of what is involved when the plumber arrives.

The plumber will test the elements with an electrical meter to see if the element is bad. This is done with the power on and bare wires exposed. For an inexperienced person, the consequences of this test could prove fatal. When it is determined the element is bad, the plumber will replace it. This is done by turning off the power and draining the tank below the level of the element.

Then the plumber will remove the old element after the wiring is disconnected from it. Plumbers have special element wrenches to remove screw-in elements. They use screwdrivers to remove bolt-in elements. They place a washer around the new element and install it in the tank. Pipe compound is applied to screw-in elements before they are installed. The wiring is replaced, and the tank is filled with water. The power is turned on, and the element is checked with the electrical meter. If it checks out, the access cover is replaced, and you will have hot water within an hour or so. This job will take a plumber less than an hour to do.

Thermostats

Most electric water heaters have two thermostats (Figure 8-5). Like the elements, thermostats require working with the 220 volt current. Again, I cannot advise you to attempt working with thermostats, but I will give you a little background information on them.

If you have no hot water and the heating elements are functioning, the next likely cause is the thermostat. Your plumber will check the thermostats with an electrical meter. If it is determined the thermostats are bad, the plumber will replace them.

This is done by turning off the power to the water heater and disconnecting the wiring. The thermostats sit close to the water tank and are buried in insulation. After the plumber replaces the thermostat, the wires are reconnected and the power is left off so the insulation can be replaced. Then the power is turned back on. The thermostat is tested with a meter, and the insulation is replaced. The access door is replaced, and the job is done. This job will take a professional plumber less than an hour.

Replacing an Electric Water Heater

If you plan to replace your own water heater, you will have to deal with the electrical wiring. For an untrained person, this is not a good idea. I will tell you how to replace the water heater, but I will not explain the wiring. Call a licensed electrician to handle the electrical work.

The act of replacing a water heater normally requires a permit from the local code enforcement office. Check on the code and legal requirements before replacing your water heater. If you have a plumber do the job for you, ask about a permit. Some plumbers don't bother to obtain a permit to replace a water heater. If the code requires a permit in your location, insist on the plumber obtaining one. The code inspection is for your protection.

The first step in replacing an electric water heater is to cut off the power and water. The next step is disconnecting the electrical service to the water heater. With the electrical work disconnected and the water turned off, drain all the water from the old heater. Cut the cold water pipe about 6 inches above the water heater. When the pipe is cut, the tank will drain faster.

When the tank is empty, cut the hot water pipe about 6 inches above the tank. The old water heater is now ready for removal. Put the new water heater in place. (This example is based on working with cop-

per pipe since it is the most common type.) In addition to the new water heater, you will need a new relief valve, copper fittings, copper pipe, pipe compound, and soldering gear. Residential water heaters are designed to work with $7/8$-inch pipe. In the trade, this is called $3/4$-inch pipe.

Apply pipe compound to the new relief valve and screw it into the tank. Refer to the manufacturer's instructions for the proper place to install the relief valve. Most heaters have a hole clearly marked for the relief valve. Some of them are on the side of the heater, and others are on the top.

Clean and flux all pipe and fittings before putting them together. Cut two pieces of pipe about 4 inches long. Some water heaters require female adapters, but most need male adapters. Solder the proper adapters onto the pipes and let the joints cool. When the adapters have cooled, apply pipe compound to the threads of the male adapters or to the nipples extending from the water heater. Connect the adapters to the connections on the heater.

Use unions, couplings, or whatever fittings are necessary to connect these risers to the hot and cold water. The water heater will have its holes marked for hot and cold water. The cold water goes to the inlet, and hot water comes from the outlet. When all the pipes are connected, solder the joints. Now you need to make a discharge tube for the relief valve if the old one cannot be used.

The discharge pipe should run from the relief valve to a point about 6 inches above the floor. This pipe is important. It reduces the risk of injury if someone is near the water heater when the relief valve blows out hot water or steam. A male adapter screws into the end of the relief valve for you to connect your copper pipe to.

When all the pipes are connected and cooled, turn on the water. Open the relief valve to allow air to escape and the tank to fill faster. When water comes out of the relief valve, close it. The last step is to have the electrical service reconnected to the water heater. When the wiring is complete and turned on, you should have hot water within two hours.

GAS-FIRED WATER HEATERS

Gas-fired water heaters (Figure 8-6) require the services of a professional for most repairs. You can work with the relief valve and the drain, but never attempt to work with the burner or the gas piping. Gas is volatile and extremely dangerous in untrained hands.

A Lack of Hot Water

If your gas-fired water heater is not producing, there are a few things you can check. Look to see if the pilot light is burning. Consult your owner's manual to see if you have electronic ignition for your burner. There may be instructions for working with the pilot mounted on the tank. If the owner's manual is not clear in what you should do, call a plumber.

Inspect your gas valve to be sure it is open. The handle should be in a straight line with the pipe. If the handle is at a right angle to the pipe, it has been cut off. Determine why it was cut off before returning it to the open position. If the valve is open, check the dial on the control box. This dial may be turned off.

This is about all the average homeowner should attempt to do with a gas-fired water heater. The possible repercussions of working with gas are too great for the inexperienced person to risk.

OIL-FIRED WATER HEATERS

Oil-fired water heaters may require the services of a licensed oil burner technician or a licensed plumber. The oil-related parts require an oil burner professional. The plumbing parts dictate a need for a plumber. You should never attempt to work on the oil equipment or burner. Limit your work to minor plumbing repairs and adjustments.

DOMESTIC HEATING COILS

Domestic heating coils (Figure 8-7) are water heating devices used in conjunction with boilers that heat your home. If your home is heated by a boiler, you probably have a domestic heating coil. These units are often called *tankless heaters* and are sometimes referred to as *instantaneous heaters.*

Domestic coils are made up of many coils of copper tubing immersed in a boiler. The potable water in the coils never comes in direct contact with the water in the boiler. This type of water heater is different from the other types described.

Coils provide instant and continuous hot water. The other styles use a storage tank and a heating device to heat the water. With the tank-style heaters, it is possible to run out of hot water. If the tank empties, it takes time for the cold water to heat up. With a coil, the water is heated as it passes through the coil. This provides constant hot water.

The drawback to a domestic coil is the amount of hot water you can receive in gallons per minute. Coils can be adjusted to produce different water temperature at different flow rates. The faster the hot water flows through the coil, the cooler it will be. If you like a strong flow of hot water, at high temperature, a coil is not your best source of hot water.

In addition to direct coils (as described above), there are *indirect coils.* These coils are not very efficient and are rarely encountered. The direct coil is immersed in the boiler. Indirect coils sit outside the boiler and do not heat the water as well.

There is little opportunity for the weekend plumber to work with domestic coils. Due to their relationship with the boiler, it is very difficult for an untrained person to work with them. If you have a coil with troubles, call a plumber.

Figure 8-7. Domestic heating coil. Courtesy of Amtrol, Inc.

9
Sump Pumps, Sewer Ejectors, and Specialty Pumps

This chapter is all about non-potable water pumps. Not all homes are equipped with these pumps, but many are. Sump pumps are common in homes with wet basements. Sewer ejectors are used to lift sewage up when the plumbing is located below the intended drainage point. The most common use of these pumps is in basement bathrooms. Other types of pumps include specialty pumps. A laundry pump is one example of a specialty pump. When a laundry tub is installed below the home's building drain, a laundry pump is used to pump the water up into the building drain. All these pumps can create problems for the homeowner. This chapter will educate you in what to do if trouble occurs.

Of all the pumps covered in this chapter, sump pumps are probably the most often encountered. The primary goal of a sump pump is to pump unwanted water to another location. A typical use is pumping water out of a basement to a suitable location outside the home.

TYPES OF SUMP PUMPS

There are two basic styles of sump pumps available. They are *submersible sump pumps* (Figure 9-1) and *vertical sump pumps* (Figure 9-2). While these are the two styles of pumps, there are many variations of each style. Some submersible sump pumps uti-

lize an internal float, and others use an external float (Figure 9-3). Vertical sump pumps use an external float.

Among professional plumbers, the submersible sump pump is usually favored (Figure 9-4). Submersible pumps with internal floats are the preference of many professionals. The location for the pump's installation may dictate a particular style. Both of these pumps are designed to drain wet basements and to transfer water. Both are simple to install and require little expertise to replace.

The standard discharge on both types of pumps is a 1¼-inch fitting. This fitting is threaded and will accept a male adapter or a male-enlarging adapter. It is common practice to increase the size of the discharge to a 1½-inch pipe size. For most homeowners, the sump pump is in a corner of the basement. The pump might be sitting in a hole in the concrete on a gravel base. But the pump should be in some type of basin.

Many homeowners place the pump in a 5-gallon plastic bucket with holes cut in it. Special basins are sold for the purpose of housing sump pumps. These basins can accept ground water and water from a perimeter drain going around the foundation of the home. The placement of the sump pump can be a cause for problems.

The Stuck Float

A float that sticks can give you two types of trouble. If the float sticks in the downward position, the pump will not cut on. If the float sticks in the upward position, the pump will not cut off. When the pump runs constantly, the motor will burn up. If the pump doesn't run when it should, the area will be flooded with water. Both of these problems can be associated with the placement of the sump pump.

Two contributing factors to this problem are the size of the pump's container and the type of discharge pipe used on the pump. Many sump pumps use a flexible hose-type pipe for the discharge. This type of flexible pipe does little to hold the pump in place. As water fills the pump's basin area, the pump can be floated and moved. If the pump moves too close to the side of the basin, the basin can cause the float to stick. Vertical sump pumps are the ones most frequently moved by water. The design and weight of vertical pumps allow easier movement than does a submersible.

The vibration from the pump's motor can also move the pump. Regardless of how the pump is moved, if the float comes into contact with and sticks on the basin, you have a problem. If your float sticks for this reason, move the pump to the center of the basin area.

To reduce the risk of this happening, you have several options. Make sure the basin area is large enough to allow adequate space for the pump and float. Use a rigid pipe for the discharge of the pump. A schedule 40 plastic drain pipe is a good choice. Place bricks around the base of the pump to limit its movement. If you want to be certain to solve the problem, replace the pump with an internal-float sump pump. When the float is inside the pump, it cannot catch on the side of the basin.

A damaged shaft can cause the float to stick. These occasions are rare when the pump is installed properly. If your float is sticking on the shaft, inspect the shaft for bent areas. Check the float stops at the top of the shaft to see if they are in good repair. These are the only common causes of a sticking float.

Electrical Problems

Average residential sump pumps run on regular household current of 115 volts. These pumps are normally plugged into an outlet rather than being hard-wired to the electrical system. Inspect the cord periodically to see that the insulation is not cut. Other than maintaining a safe cord, there is nothing the average homeowner can do for the sump pump's electrical system. Except for the pump's motor, there is little chance of an electrical malfunction. If the motor is bad, you should replace the pump.

Jammed Impellers

When sump pumps are not housed in a clean basin, they can suck up sediment and rocks. If the pump motor is running but water is not being properly pumped, you may have a jammed impeller. The impellers are located in the base of the pump. Never work on the impellers with the pump plugged in. When you suspect a jammed impeller, you have two options.

With the pump's base submerged in water, shake the pump. This shaking action may dislodge the foreign object and free the impeller. If this doesn't work, you will have to remove the pump from the basin. Refer to the section on pump replacement for instructions on how to remove the pump. Unplug the electrical cord to the pump. With the pump out, invert it and inspect the impellers. Many pumps have a strainer to protect the impellers. You may have to remove the screws at the base to gain access to the impellers. If there are no visual signs for the cause of the jamming, you probably need a new pump.

The Pump that Wouldn't Stop

There are times when your sump pump may baffle you. The pump may pump on an endless cycle. This type of problem is often misunderstood. The cause for this mysterious action is frequently a bad check valve. The check valve is the device installed in the discharge pipe to prevent water from running back down the pipe into the pump basin (Figure 9-5).

If the check valve fails, water will flow back into the pump basin, forcing the pump to continue running. This could go on for days. The pump will empty the basin, but the water runs down the pipe and refills the basin. This is a never-ending cycle until the pump motor burns up. While this can be a serious problem, it is easy to correct.

Replacing the Check Valve

The check valve may be installed with male adapters or a rubber-type coupling and stainless steel clamps. To replace the check valve, unplug the sump pump. The type with a flexible coupling is easy to replace. Loosen the clamps and remove the old check valve. Slide the new check valve into place and tighten the clamps. Some check valves use a compression nut and washer to seal themselves. These check valves can be removed by loosening the two nuts at each end of the valve body.

If your check valve is installed with male adapters, and without a union, you will have to cut the pipe. Loosen the union, if you have one, and unscrew the old check valve. Install the appropriate adapters in the new check valve. Use pipe compound on all threaded fittings. Put the valve in place and tighten the union. If you must cut the pipe, install the new check valve using couplings or unions, the proper adapters, and the necessary replacement pipe.

Replacing a Sump Pump

This is normally a very easy plumbing job. Unplug the existing pump. Disconnect the discharge pipe. This can be done by loosening unions, if they are present, or cutting the pipe. Pull the old pump out of the basin.

Place the new pump in the basin. Be sure there is adequate room for the float to move up and down. Secure the pump to ensure that it will not move around during pumping operations. Placing bricks on or around the base is a typical method of movement control. Use pipe compound on all threaded fittings. The discharge connection of the pump may require a special fitting to connect to the existing discharge pipe. The standard discharge size for sump pumps is 1¼ inches in diameter. Many plumbers install a 1½-inch pipe on the pump. Making this conversion only requires the use of a single fitting. Screw a fitting with 1¼-inch threads into the discharge outlet that will accept a 1½-inch pipe in the other end.

If your sump pump does not have a check valve in the discharge line, install one. The check valve should be installed near the pump. Connect the new pump's discharge pipe to the existing discharge pipe. It is a good idea to make one of the couplings a union. This will make future replacements much easier. The last step is to plug the pump's electrical cord into an approved electrical outlet.

With the new pump installed, fill the basin with water to test your installation. When the water is high enough to raise the pump's float, the pump should cut on. As always, follow the manufacturer's recommendations in the installation and upkeep of all equipment.

EFFLUENT PUMPS

Effluent pumps operate on the same principle as sump pumps, but there are some differences. An effluent pump is the sump pump's heavy-duty big brother (Figure 9-6). It looks very much like a submersible sump pump. The discharge of effluent pumps is usually 1½ inches or 2 inches in diameter. Some of these strong pumps are wired for 115 volts, and others are wired for 230 volts.

Effluent pumps are used for the same purposes as sump pumps but for heavier demands. They can pump more, stronger, and farther than a standard sump pump. The repair and replacement of effluent pumps is essentially the same as the examples given for sump pumps.

SEWAGE-EJECTOR PUMPS

Sewage-ejector pumps are used in two primary places in the average home. The pump may be located in a basin below the basement floor, or it may be in a pump station outside the home. These

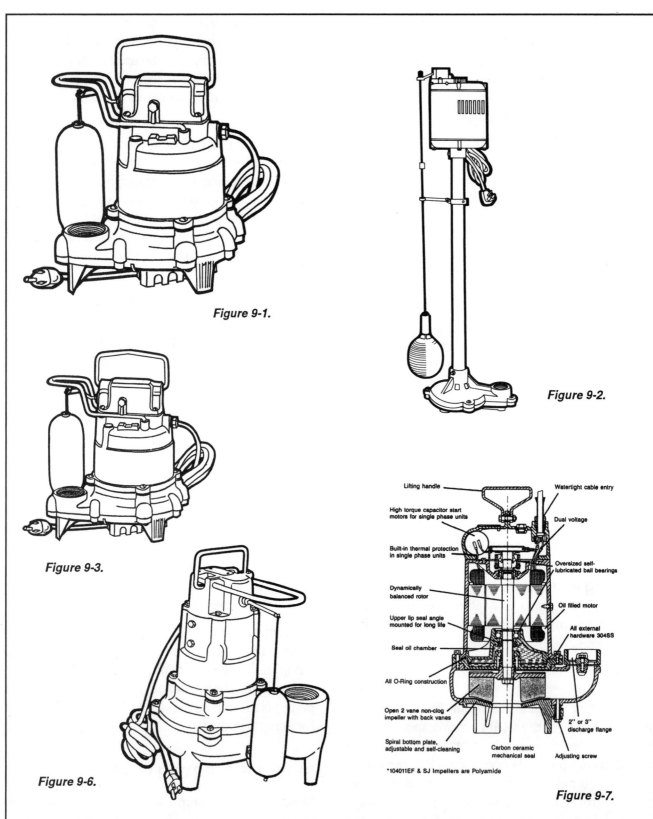

Figure 9-1.

Figure 9-2.

Figure 9-3.

Figure 9-6.

Lifting handle

Watertight cable entry

High torque capacitor start
motors for single phase units

Dual voltage

Built-in thermal protection
in single phase units

Oversized self-
lubricated ball bearings

Dynamically
balanced rotor

Oil filled motor

Upper lip seal angle
mounted for long life

All external
hardware 304SS

Seal oil chamber

All O-Ring construction

Open 2 vane non-clog
impeller with back vanes

2" or 3"
discharge flange

Spiral bottom plate,
adjustable and self-cleaning

Carbon ceramic
mechanical seal

Adjusting screw

*104011EF & SJ Impellers are Polyamide

Figure 9-7.

Figure 9-1. *Submersible sump pump. Courtesy of A.Y. MacDonald Mfg. Co.* **Figure 9-2.** *Vertical sump pump. Courtesy of Zoeller Company.* **Figure 9-3.** *External float sump pump. Courtesy of A.Y. MacDonald Mfg. Co.* **Figure 9-6.** *Effluent pump. Courtesy of Zoeller Company.* **Figure 9-7.** *Cutaway view of sewer pump. Courtesy of A. Y. MacDonald Mfg. Co.*

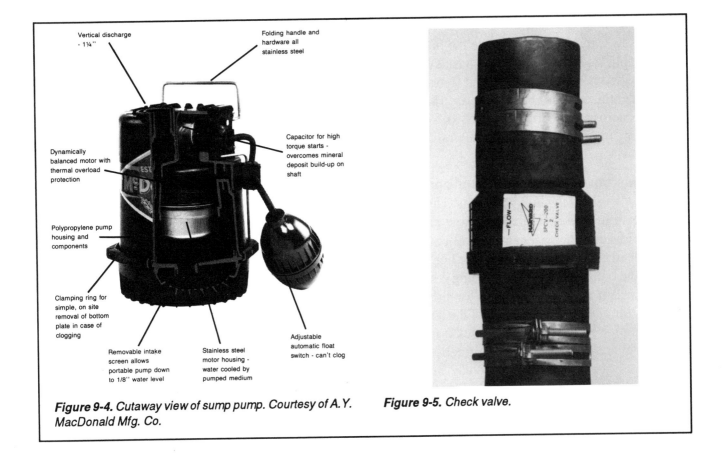

Figure 9-4. Cutaway view of sump pump. Courtesy of A.Y. MacDonald Mfg. Co.

Figure 9-5. Check valve.

pumps are used to move raw sewage to an appropriate disposal site. The destination of the pump's discharge may be a septic system, a municipal sewer, or a pipe with a gravity fall to the final destination.

When a sewage-ejector pump fails, the results can be very unpleasant. The contents of the basin or pump station will normally back up into the home, causing an awful mess. Sewage-pump stations should be equipped with an alarm system to reduce the risk of backup flooding. While most pump stations do incorporate an alarm, few basin installations offer this feature. The first sign of trouble with a basin system can be raw sewage in your home.

The sewage pump may only be used with a basement bathroom, or it may handle the sewage for the entire house. When these pumps fail to function, time can be of the essence in making the necessary repairs. The following information will give you a broad insight into the workings of sewage pumps.

The Parts of a Sewage Pump

Sewer pumps are made to be submersed in a basin or tank. They contain impellers, a shaft, a motor, upper and lower bearings, and seals (Figure 9-7). The pumps are equipped with an electrical cord to plug into an approved electrical outlet. The casing on the pump is usually cast iron. Sewer pumps are available with different horsepower ratings for various applications.

The discharge fitting is normally a 2-inch diameter, capable of handling solids up to $1\frac{1}{2}$ inches to 2 inches in diameter. The pumps are made to work with 115 volts and 230 volts, depending on the model. Sewer pumps are sometimes called *grinder pumps*. They derive this name from the nature of their operation. The pump can grind solids to a size the pump can discharge. This does not mean they can grind all foreign objects. The pumps are designed to work with normal household sewage, not items improperly entered into the system.

Sewage-Pump Basins

When a sewer pump is used inside the home, it is normally placed in a basin. These basins are typically installed beneath the basement floor. A standard residential basin is about 3 feet deep and about 30 inches across at the top (Figure 9-8). The basin has a 4-inch inlet hole to accept the plumbing drain pipe.

The basin is covered with a top, sitting on a foam seal. The top is held in place with bolts. The top has two holes in it. One is for the discharge pipe from the pump and the other is for an air vent. The air vent should ideally go up and penetrate the home's roof line without tying into another plumbing vent. While this is ideal, it is not uncommon for the vent to connect to another vent in the attic of the home. The discharge pipe from the pump comes through the basin cover and continues to its destination. When these pipes penetrate the cover, they are either sealed with a supplied gasket or are in the form of screw joints. The basin acts as a small holding tank until the amount of sewage is enough to warrant the pump's action. When the pump's float is raised, the pump will pump the contents to another location.

How Sewer Pumps in Basins are Piped

A schedule 40 male adapter is threaded into the discharge fitting on the pump. If the discharge fitting has less than a 2-inch diameter, an adapter should be used to increase the discharge pipe to a 2-inch pipe. The pipe should come up through the basin cover and be connected to a union. After the union, there should be a short piece of pipe and then a check valve. After the check valve, there should be another union and then a 2-inch gate valve. The discharge pipe should leave the gate valve and proceed to its destination. The vent pipe should connect to the cover and extend to its termination point.

Pump Stations

A pump station is normally a concrete box buried outside the home. This box houses the pump and collects the sewage. When the amount of sewage is sufficient, the pump will cut on and move it to the desired location. The discharge pipe in a pump station is usually underground, and there is no vent pipe on a pump station.

Things You Can Do to Correct Sewer Pump Problems

Ordinarily, sewer pumps do not require much attention. Unless foreign objects enter the plumbing system, the pump will normally work fine. The two most common causes of sewer pump failure are jammed impellers and stuck floats. There are times when the controls or the motor will have problems, but these defects require the attention of a professional technician.

Stuck Floats

The floats on sewer pumps can be compared to those on the previously discussed sump pumps. The principle of the float system is the same. Most pumps you encounter have an external float. The float may move up and down on a shaft, or it may move on an arm extending from the pump. If these floats become stuck, the pump will not work properly.

When the float is stuck in the upward position, the pump will not cut off. If the float is stuck in the downward position, the pump will not cut on. If your pump acts up, check the float to see if it is working according to design.

Faulty Check Valves

Just as with sump pumps, if the check valve is bad, your sewer pump will run endlessly to empty the basin. When your pump is cycling on and off with a regular rhythm, inspect the check valve. Replacing the check valve is similar to replacing the check valve in a sump pump. Close the gate valve on the discharge line and unplug the sewer pump. Be prepared to get a little messy. From this point on, the replacement procedure is the same as described in the sump pump category.

Replacing the Sewer Pump

Again, replacing a sewer pump is very much like replacing a sump pump, except it is a bit messier.

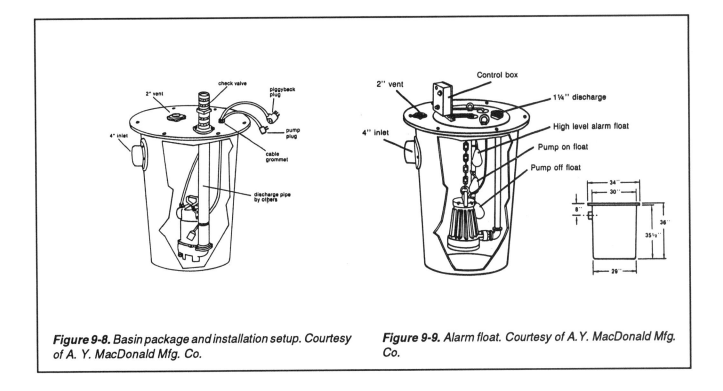

Figure 9-8. *Basin package and installation setup. Courtesy of A. Y. MacDonald Mfg. Co.*

Figure 9-9. *Alarm float. Courtesy of A.Y. MacDonald Mfg. Co.*

Before replacing the pump, close the gate valve on the discharge pipe. Unplug the existing pump from the electrical outlet. Cut the vent pipe and unbolt the cover to the basin. Loosen the union closest to the basin cover. Lift the cover over the discharge pipe. Remove the pump.

When making the pump replacement, apply pipe compound to all threaded fittings. Screw a male adapter into the discharge-outlet fitting of the pump. Install a piece of discharge pipe into the adapter. The pipe should be long enough to extend above the cover when the cover is replaced. Set the pump in the basin and replace the basin cover. Feed the electrical cord through the cover before sealing the cover. Install a check valve on the discharge pipe and connect the existing discharge pipe to the new installation.

Reconnect the vent pipe with couplings. Rubber couplings with stainless steel clamps make the job easy. When all the plumbing has been connected, plug the pump into an approved outlet. When the pipe joints are ready to be tested, fill the basin with water. You can do this by using any of the plumbing fixtures that discharge into the basin. Observe the basin and discharge pipe. When the water becomes high enough, the new pump should cut on. Normally, you will be able to hear the pump running, and you may see the discharge pipe vibrating. If in doubt, place your hand on the discharge pipe. You will be able to feel the force of the pump pushing the water up the pipe. That's all there is to replacing the average sewer pump.

Pump Alarm Systems

If your sewer pump is equipped with an alarm system, don't attempt to work on it yourself. If you are having trouble with the alarm system, call in a professional. The only exception to this rule is the inspection of the alarm float. If your alarm is sounding false alerts, check the alarm float.

The alarm float is mounted above the normal pump float (Figure 9-9). Like the other floats discussed, the alarm float can become stuck. If the float sticks in the downward position, the alarm will not function. If the float sticks in the upward position, the alarm will not cut off. This is the only aspect of the alarm system the average homeowner should attempt to work on.

SINK PUMPS

Sometimes sinks or laundry tubs are installed below the home's drainage system. When this is the case, a sink pump can be used to pump the sink's waste water up to the drainage pipe. These pumps rarely have to lift the water very high. Since the lift demands on the pump are light, the pump can be relatively small.

Direct-Mount Sink Pumps

Direct-mount sink pumps mount directly to the drain fitting of the sink. This type of pump is frequently used on basement bar sinks and laundry tubs. There are no parts of these pumps the average homeowner can work with, but there are associated fittings you can work on.

The direct-mount pump screws onto the threads of the sink drain. Water runs into the pump and is pumped out of a small discharge pipe. The discharge pipe is usually 1 inch in diameter. This pipe should have a check valve in it to prevent water running back into the pump.

There are two styles of direct-mount sink pumps. The first type is manually operated. This type requires the user to turn on a switch every time the pump is run. This can be all right in some circumstances, but the limitations prevent some applications. For example, if a washing machine discharges into a laundry tub with this type of pump, you would have to stand by to turn the pump on and off. This is not practical, and better options are available.

The other type of direct-mount sink pump has a switch device to turn the pump on automatically. The switch gauges the amount of water in or above the pump and turns the pump on at the proper time to remove the water. This style is a little more expensive, but it is much more convenient. Both styles connect to the sink in the same way and are piped to the drainage in the same manner.

Common Problems with Direct-Mount Sink Pumps

The two most common problems with direct-mount sink pumps involve the check valve and what goes down the drain. The single largest problem is what goes down the drain. These pumps are designed to pump water, not hard objects.

When these pumps are mounted on laundry tubs, they often see abuse. If a washing machine discharges into the laundry tub, lint gets into the pump. When this continues for an extended time, the lint can foul the pump. If the lint builds up enough, the pump will jam and burn up the motor. When this happens, the pump needs to be replaced.

Sand and grit are the other main killers of direct-mount sink pumps. Deep laundry sinks are used to clean everything from boots to the family pet. With so many things being washed in the sink, there is little wonder strange objects go down the drain. If pebbles or grit get into the pump, the pump can be ruined. The best way to protect the pump from unwanted objects going down the drain is the use of a strainer.

The strainer can be a basket strainer designed to fit in the sink's drain or a piece of screen wire placed over the drain. This is the easiest way to avoid a burned pump motor and expensive service calls.

The check valve can also be subject to failure from foreign objects passing through the drain. Most plumbers will enlarge the pump's discharge pipe to a 1½-inch pipe. When this is done and the check valve has a 1½-inch diameter, you will rarely have a problem. The problem arises when the discharge pipe is left in its small size and is equipped with a small check valve. Under these circumstances, the check valve can become stuck open when a hard object lodges in it. If the check valve sticks open, water can run back to the pump and cause it to cycle on and off regularly.

Replacing Direct-Mount Sink Pumps

Direct-mount sink pumps are not difficult to replace. The first step is to unplug the electrical cord from the outlet. In some cases, the electrical connection may be hard-wired into an electrical box. If your pump doesn't have a standard outlet cord, call a licensed electrician to disconnect the wiring.

If you have unions on the discharge pipe of your pump, loosen them. Ideally, these unions are between the pump and the check valve. If there are no unions, cut the pipe between the pump and the check valve. The check valve will hold back most of the water in the discharge pipe. If you disconnect the pipe between the check valve and the drainage system, expect to get wet. It is always best to disconnect the pipe between the pump and the check valve.

When the discharge pipe is disconnected, remove the discharge pipe from the pump. This is usually done by turning the pipe or an adapter counterclockwise. Next, turn the entire pump counterclockwise. It should unscrew from the sink drain without much trouble. When the pump comes off, clean the threads of the sink drain. This can be done with a rag or a brush. Remove any old pipe compound from the threads of the sink drain.

Apply new pipe compound to the threads of the sink drain. Place the new pump on the threads and turn it clockwise. Be careful not to cross-thread the fittings. Tighten the pump onto the threads. Apply pipe compound to the male adapter to be inserted in the pump's discharge outlet. Screw the adapter into the pump and connect the existing discharge pipe to your adapter. Plug the electrical cord into an approved outlet, and you are in business.

Basin-Style Sink Pumps

Some sinks are equipped with a basin-style sink pump. These pumps are the equivalent of a submersible sump pump. The repair and replacement of these pumps is the same as for a submersible sump pump. The only difference is the basin's vent. This is an air vent like the ones used with sewer-ejector basins. The air vent and discharge of these pumps are usually piped with $1\frac{1}{2}$-inch pipe.

With basin-style pumps, the sink drain is piped to the basin. From there, the pump sitting in the basin pumps the waste water to the main drainage system. The vent extends up and ties into another plumbing vent or extends through the property's roof. The suggestions given for submersible sump and sewer pumps apply to basin-style sink pumps. The floats on the pumps can be external or internal. The electrical hookup is usually a 115-volt electrical cord with a standard outlet plug.

This concludes the majority of pumps found inside the home, except for potable water pumps. Potable water pumps are covered in Chapter 15.

10
Potable Water Filters and Conditioners

Water filters and conditioners are common features of many homes. Without these units, the home's water would be less than desirable. Water treatment devices range from the very simple to the highly complex. The device may be only a small fiber filter in a canister that hangs from a water pipe. It can be an imposing system of equipment requiring several square feet of floor space. In either case, the device is used to improve the water quality.

WATER SOFTENERS

Water softeners are used to eliminate hard water (Figure 10-1). Hard water contains calcium and magnesium. These minerals cause a scale to build up on the interior of water pipes and water heaters. They can also cause a film to be left on dishes or clothes washed in the hard water. Water softeners are also used to remove small to moderate amounts of iron. When your water has iron in it, the water may be discolored. Iron-filled water can stain clothes and plumbing fixtures.

Water softeners go through four stages of operation. The first stage is called the *backwash cycle*. After the backwash, the system is put into a *regeneration phase*. A *rinse cycle* follows the regeneration. When the rinse is complete, the system returns to the service-ready mode.

These cycles vary with different manufacturers'

equipment. Refer to your owner's manual for the recommended settings for these cycles. If you decide to work on your own water softening equipment, always follow the manufacturer's guidelines. The only regular maintenance required with water softeners is adding salt to the brine tank. Whether you are setting the timer for the backwash cycle or adding salt to the brine tank, refer to the owner's manual. With the wide variety of equipment produced, it is difficult to give specific instructions on a generic basis.

The installation and replacement of water softeners should be left to professionals.

FILTERS

Filters are sold in several shapes and sizes. They can be purchased to mount on a faucet's aerator or to mount in-line on the water pipe. These filters are user-friendly and easily worked on by the average homeowner. Carbon filters are used to reduce disagreeable tastes and odors in water.

The only routine repairs needed for these filters is the replacement of the filter inside the filter housing. This is usually as simple as it appears. Cut off the water to the filter. Unscrew the filter housing and remove the used filter. Replace the old filter with a new one and screw the housing back together.

Replacing In-Line Filters

There is normally no need to replace the housing of in-line filters. If you do need to replace your housing, it is not very difficult. Cut off the water to the pipe running water through the filter. Many filters have shut-off valves close to the filter housing. Some have a valve on both sides of the housing. If you have these valves, cut them off.

There may be union connectors between the housing and the water pipe. If there are, take them apart and the housing will be removed. If you don't have unions, cut the pipe at each end of the housing to remove the unit.

The new filter housing normally has female threads at each end of it. There should be an arrow stamped into the side of the housing indicating the direction of flow for the water. Trying to solder pipe in a male adapter that is screwed into the plastic housing will cause trouble. The heat from the torch will transfer down the copper pipe and may melt the plastic housing. To avoid this, solder the male adapters to lengths of pipe before installing them.

It may make your life simpler later if you use union couplings to mate your new pipe to the existing water line. When your male adapters cool, apply pipe compound to the threads. Screw the male adapters into the filter housing. Be careful, as the plastic threads of the housing will be damaged quickly if you cross-thread the connection. When the adapters are hand-tight in the housing, finish tightening them with your adjustable wrench.

Hold the unit in place and take measurements for the last two pieces of pipe to complete the installation. Cut the pipe and fit it between the new pipe coming out of the filter housing and the old pipe. Solder these fittings and let them cool. Turn the water back on and check for leaks. That is all there is to replacing an in-line filter housing.

ACID NEUTRALIZERS

Acid neutralizers do exactly what the name implies — they neutralize the acid in your water. Having a high acid level in potable water can cause several problems. The acid will cause your pipes and plumbing equipment to deteriorate. It can cause copper pipe to develop pinhole leaks, and it can cause toilet tank parts to fall apart. The acid produces greenish stains on metallic plumbing parts.

In extreme cases, the acid may noticeably affect your health. I had a customer who was experiencing stomach problems. She could not understand what was causing her discomfort. A visit to the doctor and a water test indicated the acid level in her drinking water was the cause. I installed an acid neutralizer, and her physical condition returned to normal.

Acid neutralizers work with either soda-ash or calcium carbonate. There is nothing the average homeowner can do to repair an acid neutralizer. If you are having problems with your unit, refer to the owner's manual. You will more than likely have to call a trained service technician for any repairs or replacement.

OTHER TREATMENT DEVICES

Sand filters are used to control floating particles in potable water. These filters reduce the silt and mud that cause cloudy water. These units are not found in many homes and require the services of a trained professional.

Chlorinators are sometimes used in household applications, but they are not common. Chlorinators are used to remove bacteria and algae. In-home units require adding normal laundry bleach to maintain the proper mixture with the water reservoir. As the homeowner, you can adjust the level of chlorination, but the rest of the equipment should be dealt with by professionals.

Iodinators are not a normal household fixture, but if you have one, employ a professional to work with it. These devices are responsible for the control of bacteria, viruses, fungi, and other hazardous substances.

There are some water-conditioning systems available to treat multiple conditions. It is possible to

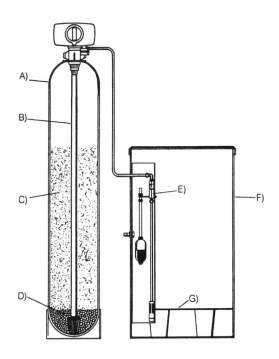

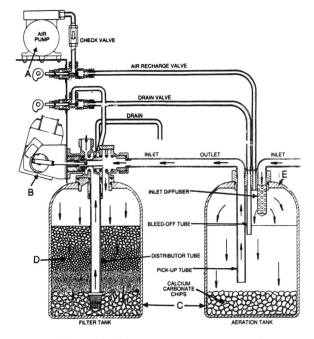

A) Durable Fiberglass Mineral Tank for rust and corrosion free service. National Sanitation Foundation Approved.

B) 1.050″ High Flow Distribution System for minimal pressure loss.

C) Hi-Capacity Cation Resin has uniform and stable resin beads assuring long life and maximum soft water. FDA Approved.

D) Gravel Underbedding assures full usage of mineral bed and allows for a smooth and constant water flow for service and re-generation.

E) Safety Brine Valve acts as a secondary shut-off in case of a power failure. In-Line Filter Screen prevents any insoluble ma-terials from being drawn into the control valve.

F) High Density Polyethylene Brine Tank has exceptional environ-mental stress crack properties. It will not crack.

G) Solid One-Piece Polypropolene Salt Grid allows insoluble ma-terials to settle out.

Parts Descriptions

A. 110 volt aeration pump automatically injects air into the aeration tank during the regeneration process without the help of injectors, venturis or micronizers.

B. Iron Curtain™ Control Center automatically controls the regeneration steps of backwash, aeration bleed-off, aeration recharge and final filter rinse before going into service. All cycles are adjustable with regeneration frequency controlled by a 12 Day Timer.

C. PolyGlass mineral tanks will not rust or corrode. National Sanitation Foundation listed.

D. Iron Curtain™ filter uses multiple layers of filtering media to provide longer filter runs, higher service flow rates and higher water clarity.

E. Iron Curtain™ Aeration System sprays raw water through a 18" head of air and then provides several minutes of additional contact time to allow the iron, manganese and/or hydrogen sulfide to fully oxidize. Special calcium carbonate chips are used to raise the pH and speed up the oxidation process.

Figure 10-1. Water softener. Courtesy of Hellenbrand Water Conditioners.

Figure 10-2. Water conditioning system. Courtesy of Hellenbrand Water Conditioners.

obtain a system to remove sulfur smells, iron tastes, and corrosion. These systems can be complex (Figure 10-2).

Replacing Complex Systems

There are so many ways for water treatment systems to be installed, it is nearly impossible to give broad-based instructions on their replacement. The other complication with these systems is their sensitivity to minor maladjustments. Water treatment is serious business and should be left to professionals in most instances.

If you insist on replacing your own equipment, follow the manufacturer's guidelines. The equipment is normally piped with standard plumbing materials. The information in this book on basic plumbing principles will usually apply to the piping of water treatment equipment. I cannot recommend that you do extensive work on your own treatment equipment. Neither can I adequately instruct you in methods to use without seeing your equipment. The best advice is to follow the instructions in your owner's manual and call a local professional if you have unanswered questions.

11
Interior Water Distribution Systems

All the water pipes in your home contribute to making up the water distribution system. These pipes are the arteries of your plumbing system. They carry water to all your plumbing fixtures. The problems you might encounter with the water distribution system are extensive. They can range from frozen pipes to water hammers. This chapter is broken down into sections dealing with the most common problems found in water distribution systems.

Your water distribution system may be composed of many different types of pipe. The pipes that are often found in the water distribution system are:

• Copper
• Polybutylene
• Galvanized steel
• CPVC
• Brass
• Polyethylene

Brass and galvanized pipe are rarely used in new plumbing systems. These pipes are found in older homes. CPVC has seen use in modern plumbing, but it is not a major contender in the materials frequently used. Copper is the most prevalent pipe used in water distribution systems. Polybutylene is a gray flexible plastic pipe and is gaining popularity as a water distribution pipe. Polyethylene is a black plastic pipe used for water services and some cold water distribution applications.

BRASS PIPE

Brass pipe is still an approved material for hot and cold water, but it is rarely used. If you are working with brass, you may think it is copper at first glance. Copper tubing cutters will cut brass, and the pipe can appear to be copper. The wall thickness of the pipe is thicker than copper, and copper fittings will not fit on brass pipe.

If you have cut a piece of pipe and cannot understand why it will not work with your copper fittings, it may be brass. Follow the pipe to the nearest fitting. If you see threads on the pipe where it meets the fitting, you can bet you have found brass. To mix brass and copper, you will need to use threaded adapters.

A common problem found with brass pipe is the deterioration of the threads. Over the years, these threaded areas become thinner and ultimately leak. The leak can be the result of pipe vibration or remodeling work. When the weakened threads are jarred, the metal can crack and leak. This problem can be overcome by replacing the damaged length of pipe and fitting with copper.

GALVANIZED STEEL PIPE

Galvanized pipe is gray and is not cut easily with common copper cutters. Galvanized pipe was a very popular water distribution pipe in its day. It is still an approved material for any above-ground applications in the water distribution system. Galvanized pipe has two major drawbacks. The threads rust through and leak, and the inside of the pipe begins to rust and catches mineral deposits. In time, galvanized pipe will close itself up with the buildup of rust and minerals. It can close to the point where the water pressure is only a trickle.

If you have easy access to galvanized water pipes, replace them when trouble comes. Repairing small sections is counter-productive when you have to continue repairing section after section. Galvanized pipe can be cut with a hacksaw or a steel pipe roller-cutter. Threaded adapters are needed to adapt other pipe types to galvanized pipe.

POLYETHYLENE PIPE

Polyethylene pipe is most often used as a water service pipe. This is the black pipe running from your well into the house or from the municipal connection to the home. Polyethylene pipe is not used with hot water, as it cannot stand the high temperature. The fittings used with this pipe are insert fittings, held in place with stainless steel clamps. Polyethylene is almost never used to convey water to individual plumbing fixtures within the home.

CPVC PIPE

CPVC is a rigid plastic pipe used for hot and cold water distribution. Many homeowners like CPVC because it does not have to be soldered. It is put together with a glue-type solvent. This pipe has met with resistance among professional plumbers. The pipe and fittings are finicky and tend to leak unless treated very gently. Professional plumbers don't enjoy the steps required to make solid CPVC connections.

The pipe and fittings should be cleaned with a cleaning fluid before installation. Then the joints are made with a glue-type solvent. The pipe and fittings must not be moved for quite some time after the connection is made. This is one reason the pros don't like it. The setting-up period slows down the job and reduces profit. The pipe becomes brittle in

cold weather. If the pipe is dropped on a hard surface, like a concrete floor, it can develop small cracks. These cracks can go unnoticed until the pressure test is applied to the system. Then the pipe leaks and requires more attention and time to correct the leaks.

CPVC is not a bad plumbing material, but you must have the time and patience required to work with it. If there is any dirt or mud on the connection, it will leak. Trying to install CPVC in the crawlspace of a home during cold weather can wear on anyone's patience.

POLYBUTYLENE PIPE

When polybutylene pipe was first introduced, it took some hard knocks. It developed a reputation for unexpected spontaneous leaks. With further development, polybutylene has captured the attention of professionals and homeowners. There are many advantages to polybutylene and very few, if any, disadvantages.

Polybutylene is a gray flexible pipe used for both hot and cold water. It can be used as a water service pipe or to pipe individual fixtures. This pipe is especially valuable in cold climates. It can expand a great deal during freezing, without splitting. The pipe does not have to be soldered; it is held in place with special clamps.

These special clamps are metal rings that are crimped around the pipe and fitting (Figure 11-1). The rings must be crimped with a special crimping tool. These tools resemble bolt cutters and can often be rented by the day (Figure 11-2). If you are uncomfortable with soldering copper pipe and fittings, I recommend the use of polybutylene.

COPPER PIPE

Copper pipe is still the mainstay of the industry when it comes to water distribution. Copper can handle hot and cold water and is touted by many professionals as the best water distribution pipe available. There are only two drawbacks to copper pipe and fittings for the homeowner. The first is cost; the second is the need for soldering skills. When compared to plastic pipe, copper seems expensive. When the cost of plastic fittings is factored in, the total cost difference may surprise you.

It is up to you to determine which type of pipe you are most comfortable working with. When you are making repairs, you may want to stick with the same type of pipe and fittings already installed. The exception to this would be when repairing galvanized or brass pipe. If you know, or can learn, how to solder, you can't go wrong with copper.

Now that you are aware of the different types of water pipes, let's explore the various problems you may encounter with them. The range of potential problems is vast. The remainder of this chapter is written to help you identify the faults in your water distribution pipes. In most cases, there are suggestions on how you can correct the problems.

PIPE CONDENSATION

Pipes in a warm environment and filled with cold water produce condensation. Many times this condensation is mistaken for a leak. If you notice droplets of water clinging to your water pipes, investigate the condensation angle. Pipes with condensation problems normally collect water over the entire pipe. This water will hang in droplets until it is ultimately displaced to a lower elevation. Wet spots on the basement floor do not always mean you have a plumbing leak. These wet spots are commonly the result of condensation.

If you have wet spots, inspect the pipes above the affected area. Wipe the pipes with a cloth and observe the pipe. Pressure leaks usually leak steadily and with some force. Condensation takes time to build up the water drops on the pipe. If the floor is marked by a long line of water stains, condensation is almost always the cause. Pressure leaks usually drip from a single location or spray water in all directions.

To reduce the effects of condensation, you can insulate the water pipes. The cold water pipe is the one you will be working with. Hot water pipes do

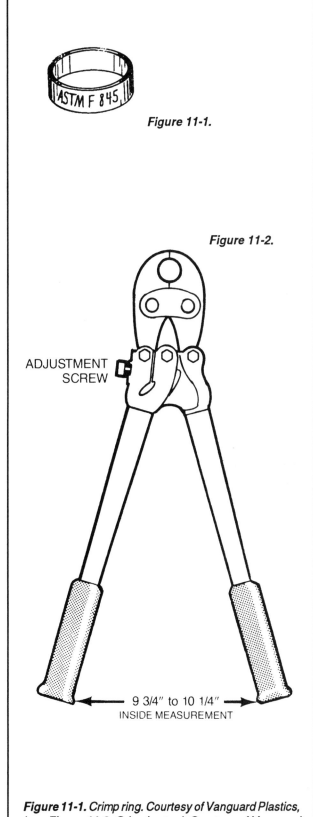

Figure 11-1.

Figure 11-2.

ADJUSTMENT SCREW

9 3/4" to 10 1/4"
INSIDE MEASUREMENT

Figure 11-1. Crimp ring. Courtesy of Vanguard Plastics, Inc. Figure 11-2. Crimping tool. Courtesy of Vanguard Plastics, Inc.

not produce condensation. Go to your local plumbing supplier and ask for a foam-type pipe insulation. You will need to know the size of the pipe the insulation will be used on. This type of insulation is sold in short lengths — 6-foot lengths are common (Figure 11-3). The cost of this pipe insulation is minimal, and the installation is a breeze.

There are two common types of pipe insulation. The first type requires you to slice it with a knife to install it on existing plumbing. Cut the insulation down its length and place it around the pipe. The cut should be rotated to the top of the pipe. Place a few wraps of duct tape on the insulation and you are done.

The other type comes already split. This type has a peel-and-stick of adhesive built into it. Place the insulation over the pipe with the split on the top of the pipe. Peel back the protective paper and join the two sides of the insulation together. Insulating the pipe will probably not prevent it from creating condensation, but it will control the amount of water dropping from the pipe.

WATER HAMMER

Banging pipes are a common complaint among homeowners. The complaint stems from the sound of a loud bang when faucets are shut off. Water hammer can also occur when a washing machine valve or a toilet's fill valve closes quickly. The loud bang that is heard is the force of water being stopped abruptly. This is most common when the water has been piped to the destination through a long, straight pipe. Undersized water pipe can also be a cause of water hammer.

If you experience water hammer, you will know it. It sounds as if someone is beating on your pipes with a hammer. There are several ways to reduce or eliminate the bothersome sound of a water hammer. The first step is to locate the fixture causing the noise. The fixture can be any faucet or valve used to dispense water.

Correcting a water hammer can involve a number of different tactics. If the pipes feeding the offending fixture are piped in a long, straight run, adding some

turns to the pipe can reduce the water hammer. This can be as simple as installing a series of 90° turns in the pipe. Another method requires adding an air chamber to the piping near the fixture. The last usual method is the installation of a water hammer arrestor.

Installing Bends to Reduce Water Hammer

Adding bends to a long run of pipe is not complicated. You cut off the water supply to the pipe. Then cut the pipe at two locations. Install an elbow to direct the pipe away from the other end of the existing pipe. Continue installing three more elbows until the pipe is no longer straight. By breaking up the straight run, the water will not develop as much force when approaching the fixture.

Adding Air Chambers

Air chambers are lengths of pipe, usually at least 12 inches tall, installed near the fixture. Air chambers should be at least 7/8 inch in diameter. To install the air chamber, you will need to install a tee fitting into the water pipe. From the tee, extend the air chamber up above the fixture and cap the pipe (Figure 11-4). The dead air space in the air chamber will cushion the effect of the water when the valve is closed. By allowing for expansion, the pipes are less likely to bang. When you combine the use of air chambers with breaking up a long, straight run of pipe, the results are usually good.

Water Hammer Arrestors

These devices are designed to deal with water hammer. The devices screw into a female thread. They should be used near and above the fixture when possible. With a permanently charged chamber and a sealed diaphragm, these devices cushion the water in the pipe. The installation of a water hammer arrestor requires cutting in a tee. The water pipe is cut, and a tee fitting is added. From the tee, there is a length of pipe with a female adapter on its end. The water hammer arrestor screws into the female adapter. You must use a pipe compound on the threads of the arrestor before screwing it into the female adapter.

FAUCET CHATTER

Water hammer noise is heard when a faucet is cut off. Faucet chatter is heard while the water is running through the faucet. This noise indicates a problem within the faucet. The most common cause of faucet chatter is a vibration within the faucet. Loose or bad faucet washers are a frequent contributor to this type of noise. Valves that are not fully open can also chatter.

Correcting faucet or valve chattering is not normally a difficult proposition. If your noise is coming from a valve, turn the handle to be sure the valve is in a full-open position. If this doesn't solve the problem, inspect the washer inside the valve. For faucets, follow the same procedure. The odds are, opening the valve or replacing the washer will solve your chatter problem.

WATER PRESSURE

Water pressure routinely stays the same unless a system is subjected to unusual events. If you have a well, water pressure is controlled at the pressure switch. This is done by the adjustment of nuts and springs, but it should not be attempted by the average homeowner. There is electrical current present at the pressure switch. If you want your pressure set up or down, call a professional.

Pressure-Reducing Valves

When a home is served by municipal water, there may be a pressure-reducing valve on the main water pipe. Pressure-reducing valves are used to control the amount of water pressure entering the water distribution system. The factory setting is normally around 50 pounds of pressure. The valve is adjustable up or down to regulate the water pressure. The adjustment is typically done by adjusting a nut and screw located on top of the valve. The nut is loosened and the screw is turned to increase or decrease the water pressure.

Before adjusting your pressure-reducing valve, refer to the manufacturer's instructions. Not all valves are the same. Allowing too much pressure into the home can have adverse effects on your faucets and pressure-relief valves. Pressure-reducing valves are an in-line type of valve. This means the water pipe comes in one end and exits at the other end.

Replacing a Pressure-Reducing Valve

Cut off the main water valve and drain the home's plumbing from the lowest spot available. If unions were used to install the valve, loosen the unions. If there are no unions, cut the pipe on both sides of the valve. This allows removal of the valve.

Apply pipe compound to two male adapters and screw them into the new valve. Install the valve so that the arrow on the valve is pointing in the direction the water is flowing through the pipe. Connect the male adapters to the existing water pipe using unions or regular couplings. When all is tight, cut the water back on and check for leaks.

BACKFLOW PREVENTERS

Backflow preventers are installed on the main water pipe going to homes served by municipal water. Not all homes have them, but backflow preventers are becoming routine in new plumbing installations. These devices protect the municipal water supply from contamination due to back-siphonage.

These devices are mounted in-line on the main water supply (Figure 11-5). There is little the average homeowner can do to service a backflow preventer. Depending on the type used, there may be some parts that a professional can repair, but upon failure, replacement is the standard procedure. If you follow the instructions given for replacing pressure-reducing valves, you can accomplish the task of replacing your backflow preventer.

THERMAL EXPANSION

Thermal expansion can drive you insane. This problem often baffles experienced plumbers. The trouble comes when a home is equipped with a backflow preventer. A symptom of thermal expansion is the release of water through a pressure-relief valve, usually at the water heater.

When a water heater's relief valve blows, it is assumed the relief valve is bad. It is not until this becomes a habitual problem that plumbers look further into the cause. The cause can be thermal expansion, caused by the installation of a backflow preventer. As the water heater heats water, the water expands. Before the backflow preventer was installed, the increased pressure from the hot water moved into the cold water pipe at the water heater.

If sufficient pressure was built up, the pressure could work down the water supply, back toward the water main. When a backflow preventer is present, the water cannot expand back to the municipal connection. With no faucets or valves open, the pressure builds until it pops the relief valve at the water heater.

Dealing with Thermal Expansion

Since you cannot remove the backflow preventer, you must take other measures. The installation of a modified air chamber is a good choice. By extending vertical air chambers over the water heater, the expanding water has somewhere to go. The air chambers should be at least 1 inch in diameter and extend as high as is reasonably possible. It may be necessary to add chambers along the water supply pipe as well. By adding enough space for the water to expand, you will eliminate the effects of thermal expansion.

CROSS CONNECTIONS

When you get hot water from a cold water pipe or cold water from a hot water pipe, you probably have a cross connection. Cross connections allow the two pipes to feed back and forth between each other. Logical locations for a cross connection are found at washing machines and faucets with spray attachments or personal shower units. These hose devices are used to deliver mixed water. When they are not protected by a backflow preventer, it is possible for the water to move from pipe to pipe through these hoses.

This is not a common problem in residential plumbing, but it does occur occasionally. If you suspect a cross connection in your system, direct attention to the devices mentioned above. Disconnect the hoses or close the valves supplying water to the hoses. If your problem ceases, you have located a cross connection. When a cross connection happens, call in a professional or replace the parts that may cause the unwanted mixing of water. Identifying the parts to replace can be tricky. Unless you are sure which part is causing the problem, call a plumber.

SADDLE VALVE LEAKS

A saddle valve is a device used to tap into a pipe. A common location for a saddle valve in the home is under the kitchen sink. If your refrigerator has an icemaker, there is a good chance you will find a saddle valve on the cold water pipe under the kitchen sink.

Saddle valves have three locations where they usually develop a leak. The first is the packing nut that holds the stem in the valve. If water is leaking around this nut, tighten the nut by turning it clockwise. If this does not stop the leak, you will have to repack the valve or replace it.

To repack the valve, cut off the water to the pipe feeding the saddle valve. Loosen the packing nut and wrap packing material around the stem. When the packing is in place, tighten the nut. Turn on the water and test for leaks. If the nut still leaks, tighten it a little more. If the leak persists, replace the valve.

Another place where leaks occur is the compression nut. This is the nut around the tubing, where it enters the saddle valve. If this nut leaks, try turning it clockwise to tighten it. If this does not work, you will have to replace the compression sleeve.

To replace the compression sleeve, or ferrule, turn the valve on the saddle valve to the off position. This is done by turning it clockwise. Loosen the compression nut and remove the tubing. Cut the tubing to remove the old ferrule. If necessary, use a compression coupling to add a new length of tubing. Slide the compression nut over the end of the tubing, with the female threads facing the saddle valve. Slide the new compression sleeve on the tubing and insert the tubing into the saddle valve. Be sure it is seated deeply and tighten the compression nut. Turn the saddle valve back on and test for leaks.

The last common place for a saddle valve to leak is the seal between the pipe and the saddle. This is usually a small piece of rubber that compresses when the saddle valve is mounted to the pipe. If water is leaking around the mounting base of the saddle valve, replace the complete valve assembly.

Replacing a Saddle Valve

Cut off the water to the pipe supplying the saddle valve. Loosen the mounting bolts; there are usually two of them. When these bolts are loose, the saddle valve will come off the pipe. Disconnect the tubing by loosening the compression nut.

To install the new saddle valve, follow the installation instructions supplied with the valve. If the valve is similar to the one removed, installation is easy. You place the mounting gasket on the saddle valve and bolt the valve to the pipe. Make sure the hole in the pipe lines up with the opening in the saddle valve. Next, insert the tubing and tighten the compression nut. Turn the handle of the saddle valve clockwise until it stops.

The next step is to turn on the water and check the mounting surface for leaks. If there are no leaks, open the valve on the saddle valve. Check the packing nut and the compression nut for leaks. If either of the nuts leaks, tighten it. If the mounting area leaks, start over and make sure the hole in the pipe lines up with the hole in the saddle valve.

FLARED CONNECTIONS

Most homes do not have flared connections, but you may run into a few. Flared connections are used for some types of gas pipe and in some mobile homes. These fittings can crack under stress and begin leaking. A flared fitting uses a fitting with threads, a retaining nut, and a flared pipe to make its connection (Figure 11-6).

When the pipe is flared, it can be weakened. This potential weak spot can crack and cause a leak. To stop a leak at a flared connection, you will have to cut the pipe and make a new flared connection. To do this, you will need a *flaring tool* (Figure 11-7). With the water off and the existing connection loose, slide the flare nut back on the pipe. Cut the pipe to remove the old flared section.

You may have to use a flare coupling to add length to the pipe for the new connection. To make a flare, insert the tubing through the proper hole in your flaring tool. The holes are labeled for the different pipe sizes. Let a small section of the pipe extend up through the flaring block. Position the flaring tool over the tubing and start turning the handle. The flaring tool will enter the pipe and roll it outward. Refer to the accompanying illustrations for the proper procedure (Figures 11-8 and 11-9). The flared portion of the pipe will fit up against the beveled part of a flare fitting. The flare nut holds the pipe in place and seals the connection.

LEAKS IN THE CEILING

Finding the source of leaks in the ceiling can be complicated. The leak may actually be several feet away from its apparent location. Just because you have a wet spot on your ceiling doesn't mean the leak is nearby. If you have water running out of your ceiling light fixture, the leak could be on the other side of the room. Water can take some strange paths before showing itself.

A pressure leak that is spraying up against the subfloor may bounce some distance before coming through the ceiling. A leak at the tub waste may run down electrical wires for several feet before showing up. If the ceiling is not level, the water may run to the lowest point to create a water stain. Much of the water's path depends on the direction your floor joists are running. When you have a ceiling leak, don't expect to find it on your first look.

Water-distribution leaks normally leak all the time. The only routine exception is a leak in the pipe feeding a shower head. This pipe only leaks when the shower is in use. Most water-pipe leaks run constantly because they are under constant pressure. These leaks are easier to locate than drainage-pipe leaks.

Figure 11-3.

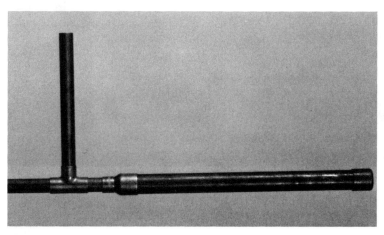

Figure 11-4.

Figure 11-5.

Figure 11-6.

Figure 11-3. *Foam pipe insulation.* **Figure 11-4.** *Air chamber on a tee.* **Figure 11-5.** *Backflow preventer.* **Figure 11-6.** *Flared fitting and flared pipe.*

Figure 11-7.

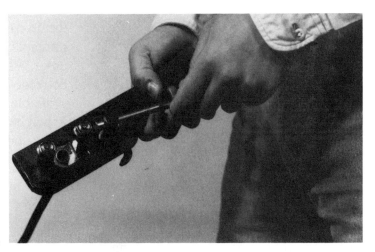

Figure 11-8.

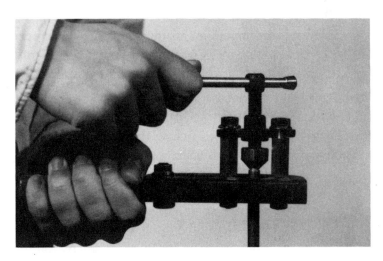

Figure 11-9.

Figure 11-7. Flaring tool. *Figure 11-8.* Pipe in flaring block. *Figure 11-9.* Flaring tool entering the pipe. Model: Andrew E. Wallace.

When you have a pressure leak in the ceiling, listen to see if you can hear the water running. If the leak is too small to hear, you will have to open the ceiling. If there are access areas above the leak, like the access panel to a bathtub, check these areas before cutting the ceiling. Finding hidden leaks is a matter of trial and error. Experienced plumbers know what to look for and where to look; you will have to experiment. With enough searching, you should be able to find a pressure leak. It is a matter of exposing enough of the plumbing to find it. This can cause significant repair bills to your ceiling. In the case of a hidden leak, it might be cheaper to call in a professional.

BROKEN AND LEAKING PIPES

To repair broken or leaking pipes, you will generally have to replace them. Follow the instructions throughout this book for the replacement of the various piping materials. Always cut off the water before attempting your repair. Normally, you will cut out the bad fitting or pipe and replace it. Following basic plumbing principles, this is not a difficult chore if you have access to the plumbing.

FROZEN PIPES AND FIXTURES

When you live in a cold climate, frozen plumbing is not a rare occurrence. There are many ways to deal with iced-up plumbing. The methods depend upon the circumstances.

There are two basic stages of frozen water pipes. There is the stage when the pipes are frozen and water cannot pass through them. Then there are pipes that split, or blow apart, and leak with abandon when the ice melts. The first is an inconvenience; the second can be an emergency.

Thawing Frozen Pipes

When your pipes are frozen, there are several methods you can employ to thaw them. The list of thawing tools is long and includes torches, hair dryers, thawing machines, heat guns, rags soaked in hot water, and a number of other remedies.

Thawing machines. Thawing machines use cables, similar to jumper cables for automotive batteries, to connect to the frozen pipe. The cables are attached to the pipe, and electrical current is used to heat conductive pipe and thaw the ice. These machines can be dangerous, and I believe they should only be used by trained professionals.

Torches. A torch will thaw frozen pipe, but it uses an open flame. If the flame gets away from you, your house could become a pile of ashes. Many plumbers use torches, but I don't recommend them for use by homeowners.

Hair dryers. Hair dryers have thawed many a pipe, but it can take a long time to accomplish the task.

Heat guns. A heat gun may be the best option for the do-it-yourselfer. A heat gun works on a similar principle as a hair dryer. It even looks like a hair dryer. A heat gun does not use an open flame, but it delivers a lot of heat to the pipe. This tool is available from rental stores and is not too expensive to purchase. In most situations, I believe a heat gun is the homeowner's best bet with frozen pipes.

Hot water. When enough hot water is applied to a frozen pipe, it may thaw. The hot water can be poured over the pipe or applied by wrapping rags soaked in hot water to the pipe. This procedure is labor-intensive and not very dependable.

Welding machines. Some professionals use welding machines to thaw pipes. This procedure can create many problems and dangers. Don't attempt to use a welding machine on your frozen pipes.

Repairing Frozen Pipes

When pipes freeze, they often split or blow apart. This requires the replacement of the damaged pipe or fitting. The method of repair varies with different types of pipe and the space available to work in. If you use soldered joints for the repair, be very careful with the torch. Frozen pipes are frequently in outside walls and surrounded by flammable materials.

If your torch sets insulation or wall sheathing on fire, it can get out of hand quickly. The wall cavity acts as a chimney and pulls the flame up. Wet these combustible materials with water and have a fire extinguisher handy. I recommend the use of compression fittings for most homeowner repairs on copper pipe. There is no fire involved, and the compression fittings make a dependable seal when properly installed. With plastic water pipe, you can cut out the bad section and replace it with the required fittings and pipe.

When a pipe is split in a tricky place, a professional can often squeeze the pipe back together and solder the split without cutting the pipe. If this procedure is not done perfectly, the result is a defective joint and more water damage down the road. If you don't have adequate access to the plumbing, or don't feel confident enough to tackle the problem, don't hesitate to call in a professional.

12
Interior Drain, Waste, and Vent Systems

This chapter is short but important. Your drain, waste, and vent system (dwv) is a critical part of your plumbing system. These are the pipes that handle and dispose of waste water and sewage. The vents allow the drains to work at their best and provide a suitable conduit for methane gas. If this part of your plumbing system breaks down, health hazards are a very real concern. In addition to health hazards, a failure in the dwv system can be quite unpleasant to the nose and other sensory equipment.

The dwv system goes largely unnoticed until it fails. The most common failure is when the drains become clogged and backed up. Advice for dealing with clogged drains is covered in the next chapter. This chapter explores preventive maintenance of the dwv system and explains how it works.

THE PARTS OF A DWV SYSTEM

To describe your dwv system, I will start at the bottom and work my way up. The main drain entering your house is the sewer. This pipe runs from the septic system or municipal sewer to the house. Once the sewer is in the home, it becomes the *building drain*. The pipes that branch off the building drain are simply *drains*. The short section of pipes running from the drain to a fixture's trap is the *trap arm*. Pipes extending above the drains and sometimes through the roof of the home are *vents*.

There are also pipes called *wet vents*. These pipes serve as a drain and vent. A common wet vent situation has a toilet that is vented by the drain of a lavatory. The lavatory drain is the wet vent, and the vertical pipe rising off the fitting at the trap arm is a dry vent.

THE PURPOSE OF DRAINAGE VENTS

Vents have two primary functions. The vent allows a plumbing drain to drain better and faster. When air is available to the drain, it can work better. This follows the same principle of punching two holes in a can of oil before pouring it into your car. The second hole introduces air and makes the filling process much faster.

Dwv vents also provide a place for sewer gas to escape. Sewer gas builds and infiltrates the plumbing system. The water seal in a fixture trap stops it from coming into the open atmosphere of the home. Vents allow the gas to be removed from the system by venting it above the roof and then into the open air.

When plumbing vents become clogged, drains don't drain as well, and sewer gas is trapped in the piping. A bird's nest and leaves are two examples of possible vent blockages. In cold climates, vents may ice up and become blocked. Your vents should stay clear of obstructions at all times.

TYPES OF DWV SYSTEMS

Homes with modern plumbing have schedule 40 plastic pipe for their dwv systems. Older homes have cast iron pipe as the primary pipe with copper or galvanized pipe used for smaller drains. In old systems, cast iron is used for the building drain and the main vent. Galvanized or copper pipe is used for drains and vents with a diameter of 2 inches or less.

The joints on this type of dwv system are caulked lead joints. This type of cast iron pipe is called *service-weight pipe*, and the fittings are *bell-and-spigot fittings*. The joints are made by pouring molten lead around the pipe to seal the joint. This is an operation that should only be performed by professionals.

The molten lead will cause severe burns if it comes into contact with your skin. Working with hot lead is very dangerous. *Never attempt to work with molten lead.* I have seen a careless plumber lose an entire body part by making a mistake with hot lead. There are other options available to you — use them.

Some homes plumbed during the transition from service-weight cast iron to plastic utilize cast iron pipe for the entire system. This type of pipe is referred to as *no-hub cast iron.* This pipe is a lighter weight than service-weight pipe. The fittings do not have a bell hub. The pipe is joined with a rubber coupling and a stainless steel clamp.

Of all these drain pipes, galvanized pipe is the most likely to cause problems. Galvanized pipe tends to rust and create rough spots inside the pipe. These rough spots catch items being drained and clog the pipe. Grease is a common offender in these cases. Galvanized drains can build up such an accumulation of grease, hair, and other particles that they will not drain at all. When this happens, the pipe should be replaced.

Even though galvanized steel is still an approved material for drainage piping, avoid using it. If you have galvanized pipe to replace, it can be replaced with schedule 40 plastic pipe. If you prefer trying to clear the blockage, refer to the next chapter. Chapter

13 discusses the various methods of dealing with stopped-up drains.

Very old plumbing systems may contain lead pipe. This pipe is usually found on lavatory and bathtub drains. Sometimes it is found where the drain pipe turns up for the toilet to connect to it. When you have problems involving lead pipe, replace it with schedule 40 plastic pipe.

Copper pipe was used for drains and vents for many years. The copper pipe has a thin wall and is called *dwv copper.* This pipe is soldered together using the same principles as in soldering copper water pipes. Dwv copper is a good pipe and rarely gives a homeowner any trouble.

CUTTING DWV PIPES

The different types of pipe require different tools and methods to cut. Plastic drains can be cut with a handsaw, a hacksaw, or roller-type plastic-pipe cutters. Galvanized pipe can be cut with a hacksaw or a roller-type steel-pipe cutter. Dwv copper can be cut with a hacksaw or a roller-type copper-pipe cutter. Lead pipe can be cut with a hacksaw. Cast iron pipe can be cut with special saw blades, but it is a very time-consuming process. Normally, cast iron is cut with a special tool, called *snap cutters* or *ratchet cutters* (Figures 12-1 and 12-2).

To cut pipe with a roller-type cutter, make sure the cutter has the proper type of wheel in it. There are different types of wheels for cutting the various types of pipe. Turn the handle of the cutter counterclockwise to open it. Fit it over the pipe and turn the handle clockwise to close the cutter. When the cutter is snug on the pipe, rotate it in circles around the pipe. As you complete rotations, tighten the handle. Continue this process until the pipe is cut.

Snap cutters and ratchet cutters are a little more difficult to use. Since you will probably be renting these tools anyway, ask the rental agent how to use the equipment. Basically, the chain with the cutting wheels is wrapped around the pipe. The chain is secured in a notch on the cutters. When the proper tension is set, the tool is ready to cut. With snap cutters, you bring the two handles together to snap-cut the pipe. With ratchet cutters, you move the handle up and down to build tension and cut the pipe. In most jobs, ratchet cutters are the easiest to use and the most effective.

There is some danger involved with cutting heavy pipe. When the pipe is cut, it, or another section of pipe, may fall on you. Always secure all sections of the piping before making any cuts.

MELDING THE OLD WITH THE NEW

When you are repairing or replacing a section of your dwv system, you may have to connect different types of pipes. There are many ways to do this, but you only need to know one way. Using rubber couplings is the easiest way to connect different types of pipes. These rubber couplings slide over the end of each pipe and are held in place by stainless steel clamps.

There is almost no occasion when these couplings cannot be used to connect a new pipe to the old one. In cases where these couplings cannot slide over the old pipe, you should call in a plumber.

DWV FITTINGS

The fittings used in a dwv system are regulated by the plumbing code. Since these codes vary from location to location, I cannot tell you with authority which fittings are legal for particular applications in your jurisdiction. Consult the local code enforcement office if you have questions about the proper type of fittings to use. The people supplying you with the fittings should be helpful and be able to explain which fittings can be used for specific applications.

THE WRAP-UP

Other than for fixture traps and clogged drains, there are few occasions when you need to work with your dwv system. These pipes last a long time and are not known for requiring regular attention. It is wise to inspect your system visually from time to time. Due to the nature of what travels through your drains, it is not healthy to let leaks go unattended. Keeping your dwv system in good working condition is necessary to ensure your good health and safety.

Figure 12-1. Snap cutters. Courtesy of Ridge Tool Company.

Figure 12-2. Ratchet cutters. Courtesy of Ridge Tool Company.

13
Drain Cleaning

There are few things in the home more annoying than backed-up drains. When a home's drains are clogged, it seems urgent to clear the blockage and get the drains running smoothly again. Professional plumbers love to be called for stopped-up drain pipes and traps. They know that in most cases the work will be easy and the money made considerable. An experienced plumber can have most drains running again in less than thirty minutes. The average plumbing company charges a minimum of one hour to respond to a service call. In addition, they usually charge a higher hourly rate to clear blocked drains.

Sometimes these extra charges are billed as an extra for equipment rental. Other companies charge extra because of the nature of the work. In reality, they charge more because they know a frustrated homeowner will pay more for this type of call. If your faucet is dripping, you might wait until the next day to get the services of a less expensive plumber. When your toilet won't flush, you will probably accept the charges of the first available plumber. Drain cleaning is a very lucrative business. It is not unusual for plumbing companies to charge between $65 and $100 per hour to clear clogged drains.

If you are willing to do the work yourself, you can save all or most of this money. At the worst, you will have to pay a nominal rental fee for drain cleaning equipment. Most interior plumbing stoppages do not require sophisticated equipment to remove them. Clogs found within the house can be cleared with inexpensive, hardware store-type equipment. These various devices should have individual prices below $20.

You can buy or rent a lot of drain cleaning equipment for the cost of a single plumbing call. Once you have purchased a plunger, closet auger, or flat-tape snake, you can use it over and over again. Acting as your own drain cleaning technician can save you hundreds of dollars during your home ownership. This chapter will instruct you in the safe, easy methods available for keeping your drains running smoothly.

KITCHEN DRAIN PROBLEMS

The first type of drain stoppages we are going to investigate is kitchen sink problems. For a simple plumbing fixture, kitchen sinks can cause a lot of trouble. They are subjected to frequent abuse and pose some challenging tests for a novice drain cleaning technician. Kitchen sinks can have one, two, or three bowls. The kitchen sink bowls are the areas that hold water. Most sinks have only one or two bowls. If there is a third bowl, it is usually designed for use with a garbage disposer.

Each bowl is equipped with a basket strainer or a garbage disposer drain. If you have a dishwasher, you may have an air-gap. Beneath the sink may be a maze of drain pipes. Many of the pipes are sensitive to movement and may leak when disturbed. These are the continuous waste and trap fittings discussed in Chapter 5.

In a single bowl sink, there are two possible drainage scenarios. In the first scenario, you have a garbage disposer. The disposer is directly attached to the sink. The water from the sink drains into the disposer and flows through it and out a small pipe on the side of the disposer. This small L-shaped pipe from the disposer is connected to the fixture trap. In the second situation, the sink is connected to the trap with a tailpiece.

In a multiple-bowl sink, the individual bowls are commonly connected by a continuous waste arrangement. These tubular pipes attach to a tailpiece from the basket strainers and may have a center or end outlet. In either case, they carry the discharge from all the bowls to a common trap. From the trap connection, the drainage pipe leaves the trap and becomes a trap arm. This trap arm continues onward to a primary drain pipe.

In most states, the trap arm enters into a stack vent. This stack vent combines a drain with a vent. The area below the connection is a drain. The pipe above the connection is a vent. The vent either extends through the roof of your home or ties into another vent pipe at a higher elevation. The drain should continue to a point where it ties into a primary drain or the building drain.

In a few states, the trap arm may continue straight to a drain with no separate vent. This combination waste-and-vent-system is rare, but allows for different tactics in clearing drain obstructions. Systems without individual back-vents can frequently be cleared with devices that will not function properly on back-vented systems. Now that you have an idea of how your kitchen drain is laid out, let's look at the type of stoppages likely to occur. Different types of stoppages require different approaches to clear them.

Kitchen sinks offer the opportunity for many types of stoppages. Foreign objects are a frequent problem with kitchen drains. Kitchen knives are a habitual offender in this category. Plastic tops from dish detergent containers are another common cause of blockages. When hard objects like these enter your drain, they are usually stopped in the trap. This is a relatively easy problem to correct.

Checking the Trap

If you have a back-vented plumbing system, you should have a "P" trap under your sink. In older homes, you may have an "S" trap. With a combination waste-and-vent system, you should have a drum trap. Any of these traps can be easily accessed. You learned how to disassemble these traps in Chapter 5. Drum traps have an easily removable clean-out plug on the bottom. A pair of t&g pliers is all that is needed to open these traps. You simply turn the clean-out plug counterclockwise, and the trap will open.

Before doing this, place a pan or bucket beneath the trap. Opening the trap allows water to escape. Once the trap is open, do a visual inspection. If you can see the cause for the obstruction, remove it. If you don't see an obvious problem, probe the accessible pipe with a piece of stiff wire (a wire coat hanger will do). With a foreign object stoppage, you will generally find the object in the trap. Once the object is removed, replace the trap and you are back in business.

When replacing the plug of a drum trap, you will need to seal the threads of the plug. You can use a Teflon® tape to seal the threads or a pipe sealing compound. Either the tape or the compound will work, but you must seal the threads. If you don't, water will leak around the unsealed threads. Slip-nuts do not require compound or tape. They operate on a compression principle.

Some slip-nut washers are rubber, and others are plastic. Inspect these washers before reinstalling them. If they are rubber and look tired, replace them. The plastic type have a long life, but the rubber type is subject to cracking and leaking when disturbed. Be sure not to cross-thread the nuts when tightening them. To test your reinstalled trap, fill the sink bowls to the flood rim with water.

Simply running water into the drain will not always expose leaks. You should fill the sink bowls and release all the water at the same time. This method increases the pressure against your resealed joints and exposes any leaks. It is a good idea to wipe the traps with a paper towel to detect minor leaks.

Liquid and Powdered Drain Openers

Another common problem with kitchen sinks is a buildup of grease in the drain line. This is a more difficult problem to resolve. This problem is intensified with older plumbing. If you have a galvanized drain pipe, you are almost certain to encounter this problem at some time. Cast iron drains are also adversely affected by greasy build-ups. At this point, let me address the use of liquid drain cleaners and powdered mixtures. These substances are dangerous and should only be used by experienced plumbers.

I know the television advertisements make these products appear to be miracle cures, but you are risking a lot to use them. They are constantly touted as being effective on hair and grease problems. In some cases, they will work, but the risk exceeds the rewards. In the first place, if they don't work, someone has to open the pipe and be exposed to the chemicals. Professional plumbers often charge an additional fee if you have filled the drain with chemicals. These additional fees range from $25 to $50.

This may not seem fair to you, but the health risks involved with chemical drain openers are significant. Before you condemn your plumber for the extra charge, consider these facts. If you have used a lye-based or acid drain-cleaning solution, the plumber could lose his eyesight. He could incur serious burns or become permanently damaged from the chemicals. If you use these solutions, always, I repeat, *always tell the plumber about the use of the chemicals*. The plumber's health and career are on the line. Your silence could ruin an individual's life (and open you to the possibility of a personal injury lawsuit).

Powdered drain cleaners rarely solve your problems; they are much more likely to add to the problem. Liquid drain cleaners available to the public are rarely effective. In certain circumstances these products will work, but there are better ways to solve the problem. Most concentrated solutions are only sold to licensed plumbers. If you get your hands on these potent solutions, be aware, your health is in danger. If you combine the use of lye and acid, you will get an eruption. The blow-back from the products can blind you.

Licensed plumbers may use a concentrated sulfuric acid to clear your drain. You should not use this solution; it is not meant for homeowner use. In general, I recommend you avoid over-the-counter, quick-fix solutions. I know from experience they will cause more trouble than they are worth. If you insist on using them, wear professional eye-protection equipment. In addition to the eye-protection, wear heavy rubber gloves. Your safety is worth much more than a call to a professional plumber. Don't gamble your future to save a few dollars.

Now that we have that out of the way, we can proceed with practical drain-cleaning methods. Metal pipe, especially galvanized pipe, has a tendency to close itself with rust and grease buildup. When these pipes are cut, you will find a mass of black unrecognizable growth extending from the walls of the pipe. When this is your problem, you will ultimately have to replace the drainage piping.

Electric Drain Openers

A drain with these characteristics can be opened today and unable to drain again in less than a month. For this condition, you will want an electric drain-cleaning machine (Figure 13-1). These devices can be rented from most equipment rental stores for a reasonable fee. A hand-held unit is easy to use and relatively safe to operate. It generally has interchangeable heads and is the best choice for metal pipes with heavy buildups.

To use this type of unit, you must first remove the trap or clean-out plug. When using any electrically operated drain machine, wear eye protection and gloves. Once you have access to the drain, feed the machine's cable into the drain. For the first cleaning, use a spring-type head (Figure 13-2). Hand-feed the cable until you come into contact with the blockage. Make sure the unit is plugged into a properly grounded electrical receptacle. Inspect the cord for any cuts or openings before using the machine. Water will generally come into contact with the cord, and a break in the cord's insulation could result in a nasty shock.

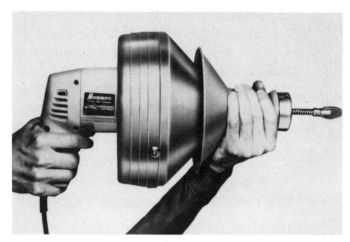

Figure 13-1.

Figure 13-3.

Figure 13-4.

Figure 13-9.

Figure 13-1. *Hand-held electric snake. Courtesy of General Wire Spring Co.* **Figure 13-3.** *Kinetic water ram. Courtesy of General Wire Spring Co.* **Figure 13-4.** *Sewer bag.* **Figure 13-9.** *Large drain-cleaning machine. Courtesy of General Wire Spring Co.*

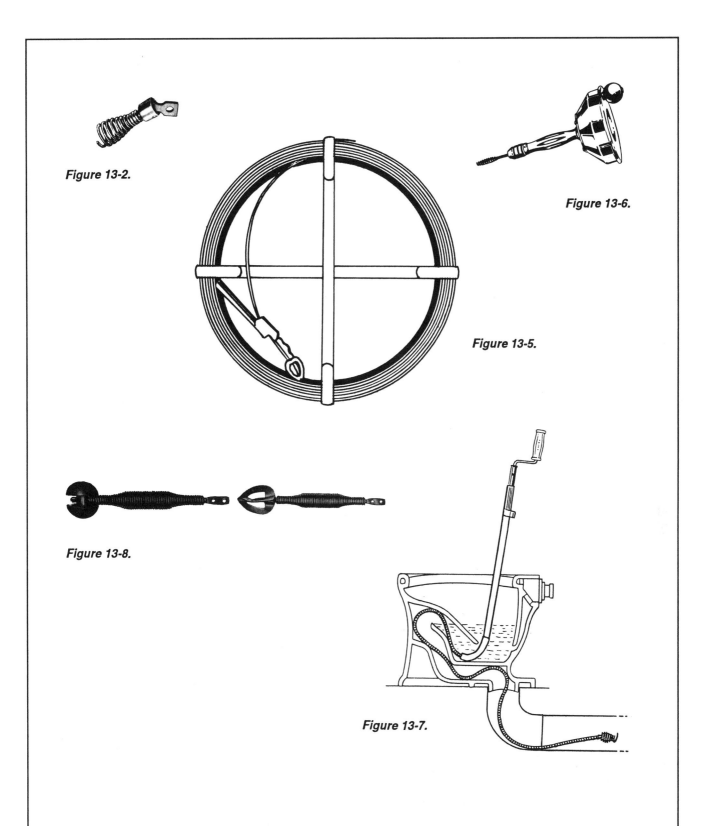

Figure 13-2.

Figure 13-6.

Figure 13-5.

Figure 13-8.

Figure 13-7.

Figure 13-2. *Spring head. Courtesy of General Wire Spring Co.* **Figure 13-5.** *Flat-tape snake. Courtesy of General Wire Spring Co.* **Figure 13-6.** *Hand-spinner snake. Courtesy of Ridge Tool Company.* **Figure 13-7.** *Closet auger. Courtesy of General Wire Spring Co.* **Figure 13-8.** *Roller-head. Courtesy of General Wire Spring Co.*

Once the blockage is encountered, make sure the machine's switch is set in the forward position. Place the nose of the machine as close to the pipe as possible. Eliminate any excess cable between the pipe and the machine. If there is slack cable, it will thrash you and make an ugly mess in your cabinet or on your floor. With the nose of the machine close to the pipe, slowly squeeze the trigger. The cable will begin to rotate. In most clogs, the cable will cut the clog or push it down the pipe.

Slowly extend the cable as the clog is moved. Some drain cleaners have automatic feeds; others require you to extend the cable manually. Follow the manufacturer's instructions in the method of extending the cable. When you have run out of cable or believe the clog is removed, extract the cable. Again, some machines allow you to do this automatically, and others require you to pull the cable back by hand. If you must remove the cable manually, use a rag to wipe the cable as it is removed. The cable will be covered with a greasy black compound.

The Test

Once the cable is removed, reconnect the drainage piping to the sink. Fill the sink and release the water all at once. Use cold water for this initial test. If the line is still blocked, you don't want the line filled with scalding hot water. Your sink should drain perfectly. If it does, repeat the process with cold water. It may take a few bowls of water to fill the pipe before the blockage is noticed.

If the drain functions properly after a few cold water tests, follow the same procedure with hot water. The hot water will help to scour the pipe and remove clinging particles. This entire process should take the inexperienced homeowner less than an hour. The procedure is simple and reasonably safe when heavy gloves and safety glasses are worn.

Plungers

If you have plastic drain pipes, you are not as likely to encounter grease blockages. They still occur but are much more rare in plastic drain lines. If you don't have a back-vented system, you may be able to clear your stoppage without disassembling the drainage piping. In a single-bowl sink, try using a plunger. With a double-bowl sink, you will need to use two plungers. Secure one plunger over a basket strainer and place the second plunger over the other basket strainer. Maintain pressure on one plunger and pump the other one.

The stationary plunger prevents air from escaping through the basket strainer. The pumping plunger builds air pressure between the plunger and the clog. Most clogs will be broken loose by this method on a combination waste-and-vent plumbing system. A kinetic water ram (Figure 13-3) performs this same action but with more pressure. These devices are harder for the homeowner to obtain, and they produce too much pressure for a double-bowl sink. The stationary plunger is not capable of containing the back-pressure from the kinetic water ram. The water ram works fine with single-bowl sinks on a combination waste-and-vent system, but is largely ineffective on double-bowl sinks.

Sewer Bags

If you have a system without back-vents, you can use a simple water-expansion rig. This device is designed to be attached to the end of a garden hose and, to be most effective, it must be placed in the drain pipe. Remove your trap and insert the sewer bag (Figure 13-4) in the drain. Insert it far enough to completely seat the device into the pipe. Once the rig is in place, turn on the water to the hose.

The device is ridged and will expand with the water pressure filling it. When the unit is fully expanded, it will send bursts of water down the pipe. The ridged surface prevents the water bag from being pushed out of the drain. These units build water pressure between themselves and the clog. In a matter of minutes, the clog should be gone. These inexpensive products are excellent when back-vents are not present. If your plumbing is back-vented, the water from the water bag will escape up the vent. In a few minutes, you will see water running off your roof if the system is back-vented. The key to using these devices effectively is getting them in direct contact with the clog. The water pressure must build between the clog and the device.

Dishwasher Drains

Occasionally a dishwasher drain will malfunction. If you believe your dishwasher drain is blocked, start your troubleshooting by disconnecting the hose from its connection to the sink drain. The hose should be connected to the garbage disposer or a wye tailpiece. Place the hose in a bucket and set the dishwasher in the drain cycle. If water flows freely from the hose, your problem is in the disposer or kitchen drain.

If water does not come out of the hose into the bucket, work your way back toward the dishwasher to find the problem. You may have to replace the drain hose from the disposer to the drain. If you determine the stoppage is in your disposer, try these steps.

Loosen the slip-nut around the disposer's drain ell and remove the lower drain piping. Remove the disposer from its mounting bracket. Once the disposer is free from the sink, turn it upside down and shake it. Before reinstalling the disposer, try running water down the fixture trap. This can be done with a garden hose or by pouring water from a pitcher.

It is important to know if the fixture's trap arm is draining properly. If it is, you know your problem is in the disposer. If the trap arm will not drain water, the problem is farther down the pipe. With this evidence, you can make an informed decision. If your trap arm drains properly, reinstall the disposer. Test the drainage and see if everything works as it should. If you still have a backed-up drain, your problem is in the disposer. At this point, you should call a professional. There is not much more to do.

Well, we have dealt with most of the problems likely to be encountered with your kitchen sink's drain problems. With obstruction blockages, you have many options. With a combination waste-and-vent system, plungers and water bags will generally solve your problem. Trap obstructions can be cleared without specialized drain-cleaning equipment. Trap-related problems can often be corrected with a pair of t&g pliers and a coat hanger.

Dishwasher hoses can be cleared or replaced with a screwdriver, t&g pliers, and possibly a piece of wire. In grease situations or closed pipes, you can rent an electric drain-cleaning machine. There are a few more useful tools available for kitchen-sink drain problems.

Other Drain Openers

Flat-tape snakes (Figure 13-5) are effective in clearing obstructions and punching holes in closed-up metal pipes. A *trap spoon* may be used to clear some trap obstructions without disassembling the trap. *Grapple-grip tools* are also effective in removing some trap obstructions without disconnecting the drainage plumbing. Another inexpensive, useful tool is the *hand-spinner snake* (Figure 13-6). This device works on a similar principle to that of a small electric drain machine. It is operated manually and does not require electricity. With a little trial-and-error experience, you should be able to solve most of your kitchen drainage problems on your own.

BATHROOM SINK CLOGS

The next area most likely to give you drainage problems is the bathroom. The bathroom sink, or lavatory, can be dealt with in much the same way as the kitchen sink. The procedures are the same, and the devices to be used are the same. Lavatory drains do not usually stall from grease buildups, but they are subject to all the other ailments common to a kitchen sink. Lavatory drains are frequently clogged with hair. This is due to daily grooming over the sink bowl and washing hair under the faucet.

Hair clogs can be cleared with any of the primary drain-cleaning tools. In some cases, the hair will be bunched up at the pop-up assembly. Before starting a full-scale attack on the drain, inspect the pop-up fittings. You may find the hair is easily removed without disturbing the trap or drain line. To inspect the pop-up assembly, try to lift the stopper device in the center of the sink out of its hole. If the piece will not separate freely, you will have to loosen the nut at the back of the pop-up assembly.

This nut may be loosened by hand or with the help of t&g pliers. When the nut is loose, the rod extending through the nut may be pulled away from the drain. This action disengages the coupling of the pop-up assembly. You should now be able to lift the pop-up plug out of the basin. Inspect it and the rod for hair buildup. A flashlight is useful in looking down the drain. If hair is found, remove it by hand or with a coat hanger.

Once this inspection is complete, place the pop-up plug back into the basin. Turn it so the slotted area is aligned with the back of the pop-up assembly. Next, push the rod you removed back into the pop-up assembly. It should go through the slot in the tail of the pop-up plug. With the rod well seated, tighten the retainer nut to lock the rod into place. Test the operation of the unit with the lift rod. If the lift rod is hard to operate, loosen the retainer nut slightly. Continue this procedure until the pop-up plug and lift rod work easily and properly. Refer to the instructions in Chapter 6 if necessary.

Plugged Toilets

Slow flushing and plugged toilets may require different tools than a sink. If the water swirls and goes down slowly, you may have a simple problem. There are two possibilities for this problem that are not drain pipe-related. The problem could be with the water level in the tank or obstructed flush holes. If the toilet will not flush at all, skip the water tank and flush hole inspections. The amount of water needed for a dependable flush varies with the age and type of toilet. Older toilets require 5 gallons of water. Newer water-saver toilets are designed to flush with 3½ gallons. The newest toilets are capable of functioning properly with only 1½ gallons.

To check the water level in your tank, remove the lid and note the level of the water. There is a line etched into the back of the tank. The line is identified with the word "fill-line." If the water is at or near this line, the tank has adequate water for a normal flush. If the water level is well below the fill-line, the ballcock needs to be adjusted. This adjustment will allow more water into the tank and may solve your problem. For instructions on adjusting the ballcock, refer to Chapter 6 in the toilet section.

If the water level is sufficient, look to the flush holes. The flush holes play an important role in the flushing of most toilets. This is an area of troubleshooting often overlooked even by experienced plumbers. Flush holes are located around the inside rim of the toilet bowl. There should be several of these holes. They are hidden from view, but are easily found with your fingers or a mirror. These openings can become clogged with mineral deposits from the water. If they are blocked, the toilet will not flush properly.

A coat hanger is the best tool for this job. Insert an end of the wire into each hole. Work the wire back and forth and side to side. If you hit resistance, you may have discovered your toilet's problem. After thoroughly reaming the holes, try flushing the toilet. If it is still slow and swirling, you must move on to drain-cleaning tools.

Start your blockage-clearing attempts with a plunger. A plunger with a large bell-shaped head is the most effective type to use on toilets. Place the plunger over the discharge hole and make several thrusts. A kinetic water ram may be used in place of a plunger. Try flushing the toilet again. If it still does not drain properly, use a closet auger (Figure 13-7). Be careful with the auger, as too much torque can crack the toilet bowl. The auger will have a large spring head to clear the stoppage. With the spring head fully retracted into the auger sleeve, place the auger in the toilet bowl. Push the handle and cable downward.

You may have to make a few attempts before getting the auger started into the toilet's internal trap. When you feel the auger's head enter the trap, start turning the auger handle clockwise. While turning the handle, continue to work the cable farther into the toilet. Most augers are equipped with a combined cleaning length of no more than 12 feet. If the auger starts to bind and twist, be gentle. Too much pressure could result in a broken toilet bowl. In a binding situation, work the handle in both directions and up and down. Try to get the cable fully extended through the toilet. When this operation is complete, remove the auger and flush the toilet. With luck, everything will work and the toilet will flush quickly.

If these steps have not corrected the problem, you will have to remove the toilet from its flange. The following procedures are applicable to all toilet drain stoppages. To remove the toilet, follow the instructions in Chapter 6, in the toilet section.

With the toilet removed, you will be looking down a 3- or 4-inch drain pipe. The best devices to open these blocked pipes are a water bag, a flat-tape snake, or an electric drain-cleaning machine. The water bag will not work if your plumbing system is back-vented. For homeowners, a flat-tape snake is one of the best choices available. Insert the head of the snake into the drain. A roller-head (Figure 13-8) is the best choice for this type of work. Push the snake into the drain as far as you can.

When you hit resistance, it could be the clog or a short-turn fitting. If it is a fitting, you should feel a solid resistance. A clog will normally have some flexibility to it. If a clog is encountered, continue to push the snake farther into the pipe. The clog should break up or become dislodged when it is pushed into a larger pipe. In some drainage systems, a flat-tape snake is incapable of negotiating the tight-turning fittings. These conditions call for a spring-cable snake.

Large Electric Drain Openers

Most spring-cable snakes large enough to clear 3- and 4-inch pipes are driven by electric drain openers (Figure 13-9). It is possible to obtain a manually-driven model, but it is an oddity among today's modern plumbing tools. The electric drain cleaners can be rented from most tool rental stores. Before going to this expense, call a few plumbers and check their prices. It might not cost much more to have a plumber do the job for you. Large drain machines are very powerful and potentially dangerous.

If you decide to tackle the job yourself, be careful. Have the tool rental representative instruct you in the proper operation of the chosen machine. Wear safety glasses and heavy gloves and avoid loose-fitting clothes. Baggy pant legs and shirt sleeves could become tangled in the spring-cable. These machines produce a lot of torque and are quite capable of doing extensive damage to the human body.

Most large machines are operated with a foot pedal. Place the machine as close to the pipe as possible. Attach a flexible spring-head and insert the cable into the drain. Push it as far as possible by hand. When you can push it no farther, make sure the machine is in the forward gear and depress the foot pedal. Follow the instructions provided with the machine and work the cable through the pipe. Be careful to avoid twisting cable caused by a solid obstacle.

If the cable tries to kink or twist, reverse the gear and back the cable out of the pipe until it is clear of the obstruction. Place the machine in forward gear and try the cleaning procedure again. If you continually hit an immovable object, you may have a broken pipe or a house trap. At this point, it may be wise to call a professional plumber. These large electric-drain cleaners are the most effective tool to use on stubborn stoppages. If you cannot clear the blockage with the use of this machine, you will probably have to rely on a professional.

Liquid drain openers are sometimes effective on drain lines with a complete blockage. These should only be used as a last resort. They seldom work and create hazardous conditions for the individual forced to work around them. If you decide to try a liquid drain cleaner, follow the instructions closely.

Bathtub Blockages

The next source of bathroom drainage problems is the bathtub. Tub drains are easy to open with the right tools. Hair is a frequent contributor to slow-draining tubs. The first step to correcting a tub problem is checking the waste-and-overflow assembly. This is the part that you operate to seal the drain of the tub and to release water standing in the bathtub. Start with the piece in the bottom of the tub that prevents water from escaping during a bath.

There are numerous types of tub wastes. Most of them allow the stopper to be lifted out of the tub. If your waste-and-overflow is operated with a handle, grasp the stopper and gently try to remove it. It should come out easily and will probably be wrapped in hair. With the stopper removed, place a large rag over the drain hole. Fill the tub partially and then remove the rag. Does the water drain quickly? If it does, you fixed it. If it doesn't, remove the face plate from the waste and overflow. The plate is usually held in place by two large screws.

When the screws are removed, you will be able to pull the plate and the attached apparatus out of the overflow pipe. When you remove this piece, it may be covered in hair. Try the filling and draining process again. If the tub still doesn't drain, you will have to use your drain-cleaning tools. A plunger is the logical place to start. First, wet a large rag and stuff it into the overflow pipe to create a sealed opening. Then place the plunger over the drain. Force air into the drain several times. Then, try the fill and drain test again.

If the problem persists, use a small spring-cable snake. The snake can be inserted through the overflow pipe unless the tub waste is connected to a drum trap. Most tubs will have a "P" trap, and the small spring-snake will pass through a "P" trap easily. When drum traps are used, access to the trap is mandatory for snaking the drain. The clean-out plug must be removed from the drum trap to allow the snake to enter the drain pipe. Drum traps are the exception rather than the rule; they are illegal in most plumbing codes.

Once the snake is in the drain line, use it in the same manner as used with sinks. The tub drain will discharge into a 1½-inch pipe. Only spring-type snakes and very small flexible-tape snakes can be used in such a small pipe. Hand-held electric sink snakes and hand-spinner snakes work well with tub drains. A common problem with older tub drains is the type of pipe used. Many old homes have galvanized pipe installed as tub drains.

The galvanized pipe will slowly close itself with rust, grease, and hair buildups. It is not unusual for these pipes to be snaked, only to stop up again in a few weeks. The snake may only punch a small hole through the buildup. After continued use, the pipe will close itself again. At some point, galvanized drains need to be replaced.

Shower stalls are a little different from bathtubs. They do not have waste-and-overflow assemblies. When the strainer is removed from the shower base, you are looking at a 2-inch drain pipe. There is typically a "P" trap connected to the shower drain. Any of the normal drain-cleaning procedures should work well with the shower drain. The only device discussed that should not be used is the large drain machine used on the toilet's drain pipe. The 2-inch drain is too small for large electric drain cleaners, but the smaller sink-cleaning machines work well.

Bidet drains can be cleaned with the tactics used on shower or sink drains. Washing machine drains and laundry sink drains can be cleared with any of the normal small drain-cleaning tools. Bar sink drains can be treated like lavatory drains. Chapter 14 contains information on other types of more unusual and larger stoppages. Chapter 14 also deals with underground drains and sewers. Root removal, collapsed sewers, and basement drains are all discussed in the next chapter.

Techniques for troubleshooting and locating stoppages are described thoroughly in Chapter 14. When the problem is more complex than the type described above, refer to Chapter 14. It deals with correcting large-scale blockages. The problems and solutions for working with limited access and unseen pipes are described. If you didn't find what you were looking for in this chapter, it's in the next one.

14
Underground Piping

Underground piping can account for some serious plumbing problems. When these pipes fail to function, there is a good chance the rest of your drainage will not work. The same can be said about your interior water distribution pipes. If the underground water pipe to your home is ailing, the rest of the water distribution system is crippled. Pipes located under concrete floors offer the same potential for plumbing disasters.

What would you do if the pipe running from your house to the sewer in the street was stopped up? Many licensed plumbers are not sure what to do with pipes they cannot see. Only experienced plumbers have dealt with enough underground problems to simplify your life. As a homeowner, it is unlikely you have any experience in locating or repairing underground pipes. This chapter will teach you what to look for and the easiest way to find it.

WATER SERVICE

The water service is the pipe running from the well or municipal water pipe to your home. This pipe is responsible for the delivery of all the potable water in your home. The depth of a water service pipe varies with geographic location. In Virginia, such a pipe may be 18 to 24 inches deep. In Maine, it may be 4 feet deep. The depth is related to the frost line.

The water service pipe must be buried deep enough not to freeze in the winter. While there are code requirements on the minimum depth, there are no requirements on the maximum depth. It is conceivable that the water service could be in a very deep trench with the sewer. Trying to locate one of these pipes with a pick and shovel can take a very long time.

If your water service breaks, how will you locate the break? You could hire a backhoe to dig up your yard until the pipe is located. This would be expensive and messy, but it would work. You have some better options available to you. If you need to locate a leak in your water service pipe, try these suggestions.

Calling the Code Enforcement Office

A call to the code enforcement office may yield helpful information. If your home is not extremely old, there may be records on file with the code office showing the location of the water service. Many jurisdictions require the plumber to file a drawing of the water service installation to obtain a permit. If this drawing is on file, it will shed light on the approximate location of the pipe.

Locating the Water Service at the House

If you have a full basement, walk around the perimeter walls and look for the water pipe coming into the house. If you have a crawlspace, crawl around the walls to locate the pipe. If your home is on a concrete slab, walk around the perimeter walls and look in all closets and cubbyholes attached to those walls for the pipe. The water service normally enters the house at one of the foundation walls. Occasionally, it will come up out of the ground near the mechanical room or water heater. With enough detective work, you should be able to locate it.

You are looking for a copper or plastic pipe about ¾ inch in diameter. When you find it, you should have some idea of where it leaves the house. This will at least give you a starting point to begin digging outside. If the pipe is copper or galvanized steel, you should be able to locate it with a quality metal detector. Most cities and towns have rental centers where metal detectors can be rented by the hour or by the day. The rental of a metal detector is a bargain compared to the hourly rates of a backhoe and operator.

Sewer and Water District Assistance

The sewer and water district should be able to tell you where the public line enters your property. A simple phone call can put you on the trail of the pipe. With the suggestions above, you should be able to find the entry point of the pipe at the house. With help from the water district, you will know the location of the pipe at the edge of your property line. If you are fortunate enough to know the starting and ending point, you can speculate with accuracy on what is between the two points.

Digging Test Pits

If all else fails, you will have to dig. Even digging has its set of trade secrets. If the water service has been leaking long, there is a good possibility there will be wet spots in the lawn. The obvious solution seems to be to dig at these wet spots. This may not be a wise decision, as water can wander in strange ways.

If you used all the suggestions given to locate your water service, the odds are high you at least know where it enters your foundation. This knowledge is very helpful when the time comes to dig. Look at the wet spots and the lay of the land. If the land is flat, the wet spots probably do represent the location of the leak. If the land slopes in either direction, the wet spots could easily be a false sign.

If you plan to replace the entire water service, test pits are unnecessary. When you only want to repair the existing pipe, test pits can save you time and money. Before you begin digging, inquire about

buried utilities and wires. Many locations have a phone number you can call to have your underground utilities marked for free. Dig your first test pit on the uphill side of the wet spot. It is logical to assume the leak may be higher than the wet spot, with the water running downhill before coming to the surface. With a few selective test pits, you should find the leak quickly.

After You Locate the Leak

When the leak is found, you will have to evaluate the best route to take in the repair. If you determine the problem is only in a small section of the pipe, repair it. This can be done by cutting out and removing only a section of the pipe. If the pipe appears to be deteriorating along its length, replace the entire water service.

Repairing an underground water service may require special fittings. With plastic pipe, the fittings are often insert fittings held in place with stainless steel clamps (Figure 14-1). With copper, the type of repair fittings depends on local code requirements. Check with your local code enforcement office for approved fittings in your area.

The type of pipe used for the repair or replacement is also subject to the approval of the code enforcement office. As an example, many locations do not allow type "M" copper to be installed below ground. Plastic pipe often must be rated at 160 psi before it can be used for an underground water service. Since these rules vary from location to location, check with your local code enforcement office for a ruling on what materials are approved.

COLLAPSED SEWERS

In old homes, it is not unusual for the sewer to break or collapse with age. Sewers made from clay and fiber-based pipes are the ones most likely to break down. Tree roots are a common cause for the dilapidation of old sewer pipes. As trees grow, so do their roots. The roots run into the pipes and may grow through the pipe or produce enough pressure to collapse the pipe. We will explore how to deal

Figure 14-1. Insert fitting in PE pipe.

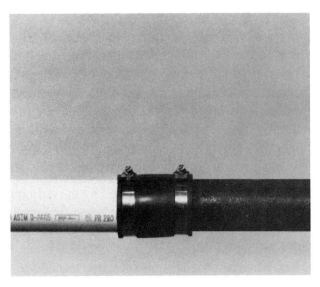

Figure 14-2. Rubber coupling for two different types of pipe.

with the roots growing into the sewer a little later. There are a number of causes for collapsed sewers, but the result is a home with slow or backed-up drains.

If you experience habitual drainage problems, your sewer may be broken or collapsed. If liquid drain openers or sewer bags are used to clear the stoppages, the sewer defect may not be noticed immediately. These drain openers may clear the stoppage at the time of use, only to have the drain stop up again later. If a small electric drain cleaner is used to clear the blockage, the broken pipe may go unnoticed. The small diameter of the snake cable may pass through the pipe without unusual effects.

When you have perpetual problems with your main drain, a large electric drain cleaner can help. By running a large cable and head down the sewer, you will find out quickly if the integrity of the pipe is suspect. This is a good time to call a professional. Using large electric drain cleaners can be dangerous. They have the power to do serious bodily damage to the operator. See the instructions for using these drain machines in Chapter 13.

When a large snake is used, corrupt pipe will twist and bind the snake cable. The torque developed when the snake encounters a broken pipe can be tremendous. The snake cable will twist and kink

around anything in its way, including the operator's arm or hand. When you expect this kind of trouble, it is best to rely on a professional for the work.

After the snake finds the collapsed pipe, you will have to dig up the pipe. To aid in finding the broken section, measure the length of the snake removed from the pipe. This will give you a good indication of where to dig when you get outside. If you had 40 feet of cable in the pipe before the snake kinked, you can use that measurement to calculate the location of the broken pipe.

Measure the distance of the pipe from the point the snake entered to where the pipe leaves the house. Deduct this from the total amount of snake inserted in the pipe to arrive at the approximate location of the break outside. Take notice of the exit location of the sewer from the house. When you get outside, look for the location of your septic tank or the municipal manhole. These reference points will provide a clue on the direction traveled by the sewer.

Sewer depth can range from 1 foot at the foundation to over 20 feet at the street connection. Average residential sewers are buried 1 and 6 feet deep. Digging up the sewer normally requires the use of a backhoe. Using your reference points and measurements, dig for the broken pipe.

What to Do When You Find It

When you find the damaged pipe, assess the situation based on the damage and the type of pipe. If the pipe is not cast iron or schedule 40 plastic, you should consider replacing the entire sewer. Spot repairs of weak sewer pipes can become a regular routine with old pipe. If the cost of replacing the sewer is extreme, you can replace the broken section and hope the remainder of the pipe stays intact.

If the damage seemed to be caused by the only tree roots in the area, a spot repair is often sufficient. If, as you expose enough pipe to replace the section, you continue to find weak pipe, replace the whole sewer. Most residential sewers today are made with schedule 40 PVC or ABS pipe. If you elect to replace the sewer, a permit will probably be required by the code enforcement office.

When you are installing schedule 40 pipe for the replacement section, you will probably need adapters to mate the different types of pipe. This is simple when you use rubber couplings (Figure 14-2). These couplings slide over each pipe and are held in place with stainless steel clamps.

INVADING TREE ROOTS

Tree roots are known for their ability to creep into the smallest opening of a pipe and strangle its passageway. Willow trees are especially marked for this behavior, but most trees have the means to impair your sewer. Roots are not much of a problem with plastic pipe, but they can wreak havoc with clay sewers. Cast iron pipe is also susceptible to root infestation. The roots generally crawl in through voids in the hub connections of these pipes.

Once the roots are inside, they seem to thrive. The roots spread along the sides of the pipe, gradually growing until they cause pipe stoppages. As toilet paper moves down the sewer, it is caught on the roots. The liquid runs past the roots, and the solids begin to build up on the root structure. This will eventually cause the drain to flow very slowly, if at all.

Like a collapsed sewer, roots may not be discovered with any drain opener except a large snake. Cutting the roots from the inside of the pipe is possible with heavy-duty drain machines and cutting heads (Figure 14-3). When using these big machines and cutting heads, danger exists from flailing cable. When the cutting head comes into contact with the roots, the snake can get out of hand. Again, for this type of drain cleaning, I recommend calling a professional. If you decide to do it yourself, be very careful.

Cutting the roots out of the pipe is a temporary solution. The roots will grow back into the pipe with time. The length of time varies, but they will be back. Unless you have an unusual case of root blockage, cutting them with a big snake is the best option for the problem.

WATER PIPE LEAKS UNDER CONCRETE

When your water pipes are under a concrete floor, a leak can be very difficult to find. There are a few indications of a leak when you can't see it. A noticeable drop in the water pressure you are accustomed to could indicate an underground leak. A well pump that cuts on at odd times when you have not been using water might mean a leak is present. Noise from water pipes above grade could tip you to an unnoticed leak.

If you suspect you have a leak under your floor, there are a few ways to be sure. Turn off all the faucets in your home. If you have a well pump, note the position of the needle on the pressure gauge. Without allowing the use of any water in the home, watch the pressure gauge of your well system. If the pressure drops, you have a leak. If you have a municipal water meter, watch the meter. If the numbers or the needle on the meter move, you have a leak.

Locating the Leak

Now you know you have a mystery leak, how will you find it? Finding the hidden leak is largely a matter of trial and error. First, be sure no water is being used in the home. Then make the home as quiet as possible. Turn off the radios and television.

Figure 14-3. Cutting head for roots. Courtesy of Ridge Tool Company.

Walk around the concrete and listen for running water. If the leak is a big one, you may hear it under the floor.

Begin your search where you anticipate plumbing might be located. This includes bathrooms, mechanical rooms, and the water service entrance location. It may also include areas where plumbing is located above on the next floor. For example, if the kitchen sink is on the floor above the concrete floor, the pipes for the sink may run under the concrete and up a wall to the sink.

If this does not reveal any clues to the leak, there are other options. Go to any water pipe you can find coming up out of the concrete. Place your hand on the pipe; you might feel vibrations from the water moving through the pipe to the leak. You can also put your ear to the pipe to see if you can hear water running.

One of these methods should expose a general location of the leak. If all these fail, there is still another possibility. Take a drinking glass and place the mouth of the glass on the concrete. Put your ear on the bottom of the glass. This technique will often allow you to hear the water running under the floor to pinpoint the leak's location.

Breaking the Concrete

Most residential slabs are 4 inches thick. To get through this concrete, you can rent an electric paving breaker or a jackhammer. Rental centers rent this equipment at reasonable rates by the hour. With these electric concrete breakers, you can open the concrete floor with minimal effort.

Fixing the Leak

To repair the leak, you will need to expose enough pipe to replace the old section and connect to good pipe. The plumbing code dictates the approved methods for making pipe connections below concrete. You should consult your local code enforcement office for the options available to you.

Closing Up the Hole in the Concrete

When the leak is fixed, backfill the hole with sand. Do not backfill with sharp stones or pieces of broken concrete. These sharp pieces may lodge against the pipe and rub a hole in it as the pipe vibrates. When the hole is backfilled, you are ready to repair the concrete. You can do this by mixing bags of sand mix with water. This creates a mortar-type of cement for repairing the floor. You can use concrete mix, but it is harder to mix and difficult to finish to a smooth surface.

These suggestions should save you hours of frustration and money if your underground plumbing gives you problems. It is not uncommon for homeowners or even plumbers to start digging at random when searching for underground plumbing. By following these guidelines, you should eliminate most of the wasted effort often associated with underground piping.

15
Well Pump Systems

When you live in areas not serviced by municipal water, you are dependent on a water pump to make your plumbing system functional. If your water pump doesn't work, you don't have any water. In rural settings, the water pump is the heart of your plumbing system. This chapter is dedicated to potable water pumps.

The water pump system can be quite complicated. In addition to the water pump, there are numerous supporting components in the system. As we discuss the different types of pumps, I will identify the important elements that go with them. In all cases, electrical current is present when working with and around water pumps. Always be aware of the electrical power and never be negligent around it.

TYPES OF WATER WELLS

Dug Wells

Dug wells are very common. These wells are not very deep, but they have a large diameter. The diameter is often 3 feet across the casing. A dug well frequently has a concrete casing to hold the water. The concrete walls limit the risk of polluted groundwater entering the water storage area. Dug wells are found where the water table is close to the top of the ground. The top of a dug well is frequently covered with a concrete cap. This cap sits on top of the concrete wall casing. These wells use a shallow-well pump located outside of the well.

Drilled Wells

Drilled wells are another very common type of well. These wells are used when the water table is deep in the earth. A drilled well should have a diameter of about 6 inches, but it is often a hundred feet deep or deeper. A steel casing is normally used to protect the water storage area of a drilled well from caving in and groundwater seepage.

Drilled wells are considered to be the best type of well available. Drilled wells have the capacity to hold a high volume of water and the ability to replenish it rapidly. This type of well typically uses a submersible pump. The pump is located in the well, suspended from a pipe and sometimes from a safety cable. In some cases, drilled wells utilize a deep-well jet pump located outside the well.

Other Types of Wells

Drilled and dug wells are the most common water wells used in today's plumbing systems. Some older homes may receive water from a *driven well*. This type of well consists of a pipe driven into the earth to a water source. Driven wells are not a standard procedure today, but they can still be found. Springs still deliver water to some homes. These types of water sources use a shallow-well pump.

WHAT DETERMINES THE TYPE OF PUMP USED?

The key to determining pump specifications is the type and depth of the well. A shallow-well pump (Figure 15-1) can be used when the water in the well is not lower than about 25 feet below the pump. When the water is between 25 and 100 feet below the pump, a deep-well jet pump (Figure 15-2) may be used. If the water level is more than a hundred feet deep, a submersible pump (Figure 15-3) is the best choice.

The recovery rate of the well is another consideration in pump sizing. The pump should never be rated for a higher water delivery than the well is capable of replenishing. Pumps are rated in gallons of water pumped per minute (gpm). A pump that is able to pump 5 gallons per minute is an average pump. But if your well will only produce 3 gallons of water per minute, you must use a lower-rated pump. When the pump is able to pump faster than the well recovers, the pump can burn itself out by pumping the well dry.

ADDITIONAL WELL SYSTEM COMPONENTS

Pressure Tanks

With any type of pump, a water storage tank is desirable. These tanks are often referred to as *pressure tanks* (Figure 15-4). The size of the tank varies with the demands of the water system. The larger the tank is, the less often the pump is required to run. This extends the life of the pump. A tank with a capacity around 40 gallons is considered an average size for residential purposes.

Pressure Switches

The pressure switch controls the operation of the pump. It tells the pump when to pump water and when to stop. Pressure switches are usually located on a tank-tee or pipe, very near the pressure tank. Pressure switches play a large role in the performance of your pump.

A standard setting for a pressure switch is a cut-in pressure of 20 pounds per square inch (psi) and a cut-out pressure of 40 psi. These settings are adjustable and may vary from system to system.

Pressure Gauges

The pressure gauge is often found on the tank-tee. It provides a source for visual inspection of the current water pressure available. In some installations, the gauge may be found mounted on a shal-

low-well pump. Standard pressure gauges read up to 100 pounds per square inch.

Air-Volume Controls

Air-volume controls (diaphragm type) maintain a set ratio of air in the pressure tank. This ratio is what maintains pressure and prevents the pressure tank from becoming waterlogged. If the air pressure is not present, the tank will fill with water. A tank filled with water will cause the pump to run each time a faucet is turned on. This defeats the purpose of having a pressure tank.

There are float-type air-volume controls, but the diaphragm control is the most common. This air-volume control introduces a limited amount of air into the system each time the pump shuts off, if it is needed. These diaphragm controls can be used with shallow-well and deep-well jet pumps.

Newer pressure tanks use a diaphragm in the tank to maintain a stable air pressure. These tanks do not require the use of an air-volume control valve. The tanks with internal diaphragms are called *captive-air tanks*. This type of tank is routine in new installations and should be used in replacements of older galvanized tanks.

Pressure-Relief Valves

These pressure-relief valves act in a similar manner to those found on water heaters. The relief valve on a pump system reacts to high pressure but not temperature. If the pump continues to pump beyond the designated pressure of the storage tank, the relief valve will blow off. This is a safety precaution to prevent the explosion of the pressure tank.

Drain-Back Systems

In cold climates, drain-back systems are sometimes used. These systems allow the water in the pipe from the well to the pump to drain back into the well. This action prevents the freezing of the supply line from the well.

Foot Valves

Foot valves (Figure 15-6) are used with systems

having a suction pipe, such as a jet pump. The foot valve doubles as a strainer and a check valve. The check valve duty holds water in the supply pipe. This eliminates the need for constant priming of the pump. The strainer prevents the influx of gravel and large particles of debris into the line. It does not, however, prevent sand from entering the supply pipe.

Check Valves

Check valves are installed in the supply pipe from the well, before it reaches the pressure tank. The check valve prevents water from siphoning back from the tank into the well. If all the water runs back into the well, the pump will generally have to be primed before it will pump water back into the tank. This is not true with drain-back systems, but it is true of the average well system.

Centrifugal Pumps

Centrifugal pumps use centrifugal force to move water. The distance the water is moved depends on the degree of centrifugal force. Impellers are used to create this force (Figure 15-7). The size and number of rotations per minute of the impeller dictate the delivery of water. In a centrifugal pump, the impeller is the only moving part.

To increase pressure in a centrifugal pump, the speed of the water must be slowed. This is done by controlling the size and number of the impellers and their speed of rotation. *Booster pumps* are also used to increase water pressure. When a pump has only one impeller, it is called a *single-stage pump*. When the pump has more than one impeller, it is a *multi-stage pump* (Figure 15-8). Centrifugal pumps are compact and simple.

Jet Pumps

Jet pumps are very common and may use a one- or two-pipe system. In a single-pipe system, the pump creates suction to pull the water from the well. In a two-pipe system, the pump forces pressure down one pipe and up the other. The single-pipe system is used on shallow wells, and the two-pipe setup is used on deeper wells, up to 100 feet deep. Jet pumps are mounted in a location outside the well casing.

Submersible Pumps

Submersible pumps are submerged in the well. Water is taken in one end of the pump and pushed out the other into the supply pipe. This action is accomplished with the use of impellers. The design of a submersible pump allows it to stay under the water level at all times. Submersible pumps have a control box located near the pressure tank. This control box is where the electrical wires for the pump are connected and controlled.

Pump Problems

The range of potential pump problems is vast. Some of the problems are directly related to electrical conditions. Some are the result of piping problems. Others lie somewhere between the plumbing and the electrical components. Troubleshooting and correcting pump failures is a specialty for some contractors. The field is large enough to allow specialization. Anytime you are working in a specialized field, the possibilities can be intimidating.

While there are aspects of your well system best left to professionals, there are plenty of opportunities for you to save money by making the repairs yourself. The information given to this point should provide a basic understanding of your pump and related equipment. The following information delves into the repair and replacement of your system. The only major area I cannot guide you through is the electrical side of the system. Many plumbers have trouble with the electrical requirements of a well system. For the average homeowner, the results of working with the electrical aspects could be regrettable.

While I will not advise you to work with any electrical wiring, I will give you a broad-based description of what an electrician or a plumber looks for. This information is only to help you understand the troubleshooting process. It is not intended as how-to instructions. Under no circumstances should you attempt to work with electrical power unless you have qualified skills in the electrical field.

ELECTRICAL TESTING

Troubleshooting a pump requires the ability to understand and work with electricity. This power may be in the form of 110 volts or 220 volts. If you are not skilled in electrical work, do not attempt working with the electrical parts of your system. The result of a mistake could be fatal. To test the electrical part of your well system effectively, you must have a working knowledge of electricity. This includes knowledge of ohms, amps, and voltage. Due to the high risk involved with electricity, I am unable to guide you in troubleshooting the electrical aspects of your well system. For the inexperienced person, electricity can be deadly.

Voltage Tests

The licensed technician you hire to check your pump's wiring may use many methods to determine the cause of your pump problems. One of these methods involves testing the voltage. The technician will test the system with the electrical power on and off. The voltage test tells the technician if the system has undersized wire or an inadequate transformer.

Testing for Amps

This test is done with the pump's motor running. After checking the amps, the technician compares the findings with the manufacturer's recommended ratings.

Checking the Ohms

This test is done with the power off. False readings cause a common problem in testing for ohms. If the leads of the meter or the bare wires touch the technician, the reading will not be accurate. The same is true if they touch any other ground. The wires must be disconnected for this test and must not touch a ground.

Continuity

A continuity test may be done to verify the condition of the pump's motor. This test exposes malfunctions with the motor windings and circuit. When the meter shows an unusually high reading, the technician will suspect an open connection. This could be a burned wire or a problem with the motor. If the reading is excessively low, the problem may be a short in the electrical system. This test can also indicate an improper wiring connection.

Broken Insulation

If the insulation on the pump's wiring becomes cracked, it can cause malfunctions in the pump. The broken insulation can cause a short in the wiring. Testing the insulation will determine if a submersible pump must be pulled out of the well. If the meter readings indicate a problem, the technician will have to check the pump's motor, wiring, and electrical connections to determine the exact cause of the defect.

REPLACING A SUBMERSIBLE PUMP

Replacing a submersible pump can require some real muscle. Older installations were piped with galvanized steel pipe. This pipe gets very heavy in long lengths. Modern installations utilize coiled plastic pipe. While the plastic pipe is much easier to work with, it is still heavy with a pump at the other end. If you have thoughts of pulling your own pump, plan on getting some help.

You may be able to rent a pump puller from the local rental center. If you have a deep well, a pump puller or some extra help is a must. The pump puller is essentially a winch and a roller to pull the plastic pipe out of the well. As the winch pulls the pipe, it is rolled onto the puller. This equipment can save a strained back and much time.

When you are ready to do the replacement, cut off all power to the pump. Cut off the main shut-off valve on the cold water pipe. Remove the cap on the well casing. Have the electrical wiring disconnected at the well. When you look down into the

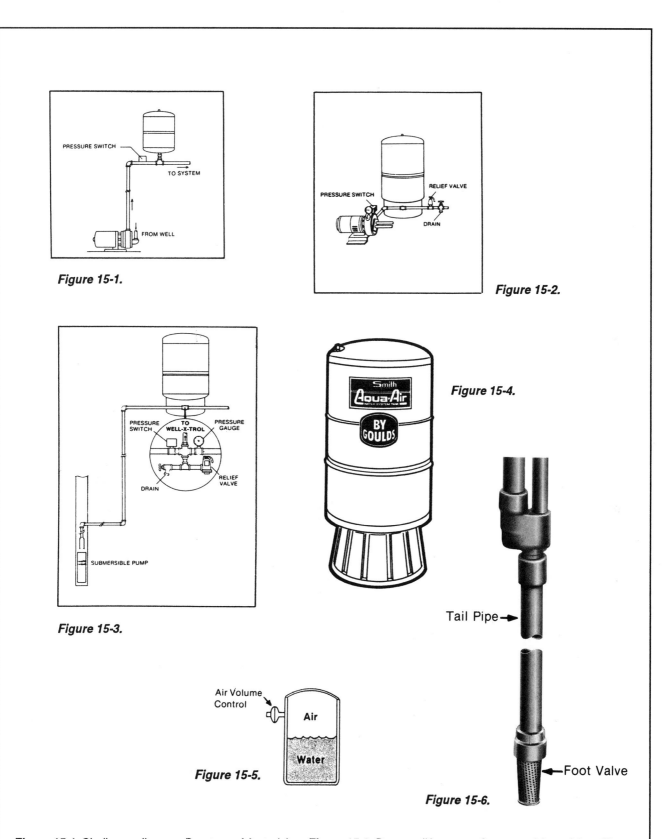

Figure 15-1.

Figure 15-2.

Figure 15-3.

Figure 15-4.

Tail Pipe→

Figure 15-5.

Air Volume Control

Air

Water

Foot Valve

Figure 15-6.

Figure 15-1. Shallow-well pump. Courtesy of Amtrol, Inc. **Figure 15-2.** *Deep-well jet pump. Courtesy of Amtrol, Inc.* **Figure 15-3.** *Submersible pump. Courtesy of Amtrol, Inc.* **Figure 15-4.** *Pressure tank. Courtesy of Goulds Pumps, Inc.* **Figure 15-5.** *Air-volume controls. Courtesy of Goulds Pumps, Inc.* **Figure 15-6.** *Foot valve. Courtesy of Goulds Pumps, Inc.*

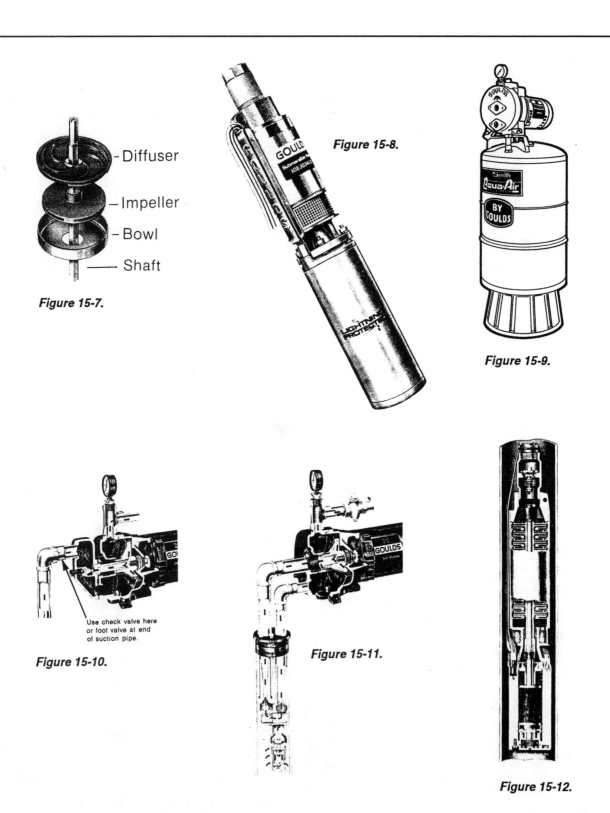

– Diffuser

– Impeller

– Bowl

– Shaft

Figure 15-7.

Figure 15-8.

Figure 15-9.

Use check valve here
or foot valve at end
of suction pipe.

Figure 15-10.

Figure 15-11.

Figure 15-12.

Figure 15-7. Impellers. Courtesy of Goulds Pumps, Inc. *Figure 15-8.* Multi-stage pump. Courtesy of Goulds Pumps, Inc. *Figure 15-9.* Pressure tank with pump bracket and pump. Courtesy of Goulds Pumps, Inc. *Figure 15-10.* Cutaway view of shallow-well jet pump. Courtesy of Goulds Pumps, Inc. *Figure 15-11.* Cutaway view of deep-well jet pump. Courtesy of Goulds Pumps, Inc. *Figure 15-12.* Cutaway view of submersible pump. Courtesy of Goulds Pumps, Inc.

well, you should see the pipe going down and a safety rope. When your water line is run under the ground, there is normally a *pitless adapter* in the side of the well casing.

The pitless adapter allows the vertical pipe in the well to connect to the horizontal pipe in the ground. To pull the pump, you may need a length of threaded pipe to screw into the pitless adapter. If the adapter already has a pipe rising to the top of the casing, you can start pulling the pump. If not, screw a length of pipe into the top of the pitless adapter to enable the pulling of the pump.

Pulling the pump is only a matter of pulling the pipe out of the well. This is fairly easy with a pump puller. Be careful not to allow the pipe or electrical wires to rub against the casing during the pulling. The casing may cause cuts in the pipe or wires. When you have the old pump out, inspect the pipe and wires. If everything is in good shape, you can attach the new pump to the existing pipe. All threaded connections should be sealed with pipe joint compound.

Submersible pumps are available as two-wire and three-wire pumps. Make sure your replacement pump is compatible with the existing wiring. Remove the old pump from the pipe. With plastic pipe, this is done by loosening the stainless steel clamps and removing the insert-by-male adapter. Most plumbers prefer to use brass insert-by-male adapters, but they are also available in nylon.

With the old pump out of the way, affix the new pump to the pipe. Follow the manufacturer's guidelines for the installation. Basically, all you have to do is reverse the pulling procedure. When the new pump is installed, the pipe may have to be cleared of debris. The pump should now be ready to run. Open the main water shut-off valve and run water through a hose bibb to clear the pipes of any sediment. If you run the water from a faucet to clear the line, sediment may clog the faucet. Most manufacturers provide concise instructions for installing their pumps.

Replacing Jet Pumps

Jet pumps are replaced using different methods than those used on submersible pumps. Jet pumps are much easier to work with. Since jet pumps are mounted outside the well, you do not have to pull the pipe up from the well. Jet pumps are normally found in basements, crawlspaces, or pump houses. Some jet pumps are mounted directly to the pressure tank.

Replacing your old jet pump is very simple if the replacement is compatible with the existing conditions. All threaded connections should be sealed with pipe joint compound. To remove the old pump, cut off the electrical power and the main water shut-off to the house. Have the electrical wiring disconnected. Loosen the stainless steel clamps around the insert-by-male adapters and remove the black plastic pipe from the well pump. Make sure your replacement pump is compatible. Your pump dealer will be able to provide you with a comparable pump when given the make and model number of the old pump.

Loosen the mounting bolts at the base of the pump. The pump is now ready to be removed. Bolt the new pump into place and reconnect the existing plumbing. Follow the manufacturer's recommendations for the new installation. Generally, it is just a matter of reversing the removal process. You may need a new pressure gauge, but if the old one is in good condition, it can be used.

When the pump is installed, have the wiring reconnected. You will normally have to prime the pump for the first running. Remove the plug on top of the pump; most are labeled as a *prime plug*. With a two-pipe system, fill both pipes with water. Pour water into the prime hole until it is full. Apply pipe compound to the threads on the plug and replace it. The pump should now be ready to run. Open the main water shut-off valve and run water through a hose bibb to clear the pipes of any sediment.

Replacing the Pressure Gauge

Replacing the pressure gauge on your well system is quite easy. The pressure gauge may screw into the pump or into a fitting near the pressure tank. Cut off the electrical power to the pump. Cut off all the water valves to and from the pump. With an adjustable wrench, turn the existing pressure gauge counterclockwise. The gauge should come out without a fuss.

Apply pipe compound to the new pressure gauge and screw it into the opening. Turn it clockwise until it is hand-tight. Use the adjustable wrench to complete the tightening process. Open the water valves and cut on the power to the pump. Open a faucet to put demand on the pump. Run the water briefly and check the gauge. If it is showing a good reading and is not leaking, you are done.

Replacing the Pressure Tank

Pressure tank replacement is not too complicated. Most pressure tanks are free-standing, independent units. Some pressure tanks are made with a bracket for mounting a jet pump. Pressure tanks are available in many sizes and styles. Some stand vertically, and others are designed to rest horizontally. The larger the capacity of the pressure tank, the less your pump will have to cycle on and off. When you use a larger tank, your pump should last longer before needing replacement.

Acquire a tank similar to the one you will be replacing. Cut off the electrical power to the pump. Turn off all water valves going to and from the tank. The next step is to drain the old tank. There should be a boiler drain on the tank-tee or in the tank. When you open this drain valve, the tank will drain. If the tank has a plug in the top of it, remove the plug. Allowing air into the tank during the draining speeds up the process. When the tank is empty, you are ready to disconnect the piping.

You will have an inlet pipe and an outlet pipe to disconnect. Most of these pipes are plastic and held in place with stainless steel clamps. Some tanks may be piped with copper pipe and soldered joints or threaded pipe and screw joints. In any case, disconnect the pipes from the tank-tee. When these pipes are removed, lay the tank on its back. Use a pipe wrench to turn the tank-tee counterclockwise. When the tank-tee is removed, discard the old tank.

There should be no reason you cannot use the old tank-tee with the new pressure tank. Apply pipe compound to the threads on the tank-tee and install it in the new tank. Set the tank in place and reconnect the inlet and outlet pipes. Modern pressure tanks come pre-charged with air to eliminate the need to pump air into the tank. When you have all the connections made, turn the water and electrical power back on. Check all connections for leaks and check the pressure on the pressure gauge. If all is well, you are done.

For tanks with a pump bracket, you will have to remove the pump to replace the tank (Figure 15-9). You can use the information given earlier about replacing jet pumps to accomplish this. When replacing a pressure tank used with a jet pump, you may have to prime the pump for the first running of the pump.

Replacing Check Valves

Check valves are also generally easy to replace. The check valve can be located anywhere in the pipe supplying water to the pressure tank from the well. Check valves are marked with an arrow on the side of the valve. This arrow indicates the direction of the water flow. Most check valves are made to accept male threads. The type of pipe used dictates the style of adapter needed for the check valve. In most well systems, insert-by-male adapters are used with plastic pipe. In copper pipe, a standard copper male adapter is used.

To replace your check valve, turn off the electrical power to the pump and close all water valves on either side of the check valve. If there is no valve between the check valve and the pressure tank, drain the tank before removing the check valve. When the old valve is installed with insert-by-male adapters, loosen the stainless steel clamps to remove the check valve. With copper pipe, cut the pipe on both sides of the check valve.

Apply pipe compound to the new adapters and screw them into the new check valve. Connect the existing pipes to the check valve using the adapters and/or couplings. When all the connections are solid, turn the water and the power back on. Run water from a hose bibb or faucet and check for leaks on the check valve.

Replacing Pressure Switches, Control Boxes, and Relief Valves

Pressure switch replacement should be left to trained professionals. The pressure switch requires working with electrical wiring for replacement.

Control boxes should be replaced by trained professionals. This type of replacement involves working with electrical wiring.

The pressure-relief valve on a well system is something you can replace yourself. The relief valve is normally found on the tank-tee. To replace the relief valve, turn off the power to the pump and shut the water valves on both sides of the relief valve. If there is no water valve between the relief valve and the pressure tank, drain the tank.

With t&g pliers or a pipe wrench, turn the relief valve counterclockwise. With a few turns, it should come right out. Make sure your new pressure relief valve carries the same rating as the one being replaced. Apply pipe compound to the threads of the new relief valve and screw it into the existing hole. Tighten the valve with pliers or a pipe wrench. Turn the water and power back on to the system. Run the water and look for leaks.

Replacing Foot Valves

To replace the foot valve, you will have to pull the pipe up out of the well. Foot valves are used with jet pumps. They are the last piece on the end of the pipe in the well.

Remove the cap from the well casing and pull up the pipe. When the pipe is out of the well, you will see the foot valve on the end of the pipe. It is a cone-shaped device screwed onto the jet body. To remove the old foot valve, simply turn it counterclockwise. Follow the manufacturer's suggestions in installing the new foot valve. Essentially, it just screws onto the jet body. When the replacement is complete, lower the pipe back into the well.

INTERNAL PARTS OF THE PUMP

Disassembly of your pump should be left to the professionals. If you are confident in your ability to undertake the task, refer to your owner's manual. There are many parts and seals inside a pump. The wrong moves can cause significant damage to the equipment. The cutaway illustrations of the jet and submersible pumps give you an idea of what is in your pump (Figures 15-10, 15-11, and 15-12). Don't rely on these examples to be the same as your pump. Always refer to your owner's manual when working with your pump.

16
Septic Systems

From a plumber's point of view, septic systems are only the place where the building sewer ends. The only involvement plumbers normally have with a septic system is when the sewage is backing up into the house. From a homeowner's perspective, the septic system is an expensive, mandatory part of the home. If you live in an area not served by a municipal sewer, you more than likely have a septic system.

Today's septic systems can be quite complex. The basic components consist of a tank, slotted pipe, and gravel. On the surface this doesn't sound like much, but if you have to pay to have it replaced, you will change your mind on its value. There is not much you need to do to maintain your septic system, but it is crucial to your home's plumbing.

When septic systems are installed, there is usually a septic plan filed with the local jurisdiction. This might be the code enforcement office or a county extension agency. If you don't have a copy of your septic design, it is wise to obtain one. If you own the property long enough, there will come a time when the septic system needs major attention. This chapter is designed to help you avoid premature expenses with your septic system.

HOW SEPTIC SYSTEMS WORK

To give you a clear understanding of why this advice is important, I am going to explain briefly how your septic system is made and how it works. When your main plumbing drain leaves the house, it goes to the septic tank. In most modern septic systems, the tank is a watertight, precast, reinforced concrete box. Older systems may have a tank made of metal, clay, or concrete block. There will be an inlet and an outlet to the box. There should be removable covers near the inlet and outlet.

These covers allow for the inspection and pumping of the tank. If you look into the access holes, you should see the inlet and outlet inverts or holes. The invert for the inlet should be about 3 inches above the liquid level. This allows the incoming waste to drop into the tank without causing a backup in the drain pipe. There is typically a 4-inch sanitary tee fitting on the drain pipe coming into the tank. The pipe going from the tee into the liquid should not go deeper than the outlet drop. Six to 12 inches is a common depth for the inlet drop to extend into the tank. On the outlet side, the drop from the tee will penetrate the liquid to a depth of about 16 inches.

The contents of the tank can be divided into three groups. There is sludge on the bottom, liquid in the middle, and a scum layer on the top of the liquid. These three layers are necessary for the system to function properly. When waste enters the tank, it is attacked by anaerobic bacteria. This bacteria breaks up the solids. Solids that are not dissolved drop to the bottom of the tank, forming the sludge layer. The scum layer is composed of solids and gases floating on the liquid. Septic tank professionals measure the sludge and scum layers to determine the need to pump the tank.

When the effluent leaves the septic tank, it goes to a distribution box through a watertight pipe. Once in the distribution box, the effluent is dispersed into the drainfield pipes. These pipes are slotted and bedded in gravel. The effluent passes through the pipe and gravel into the soil. During this phase, aerobic bacteria finish the job of decomposing the effluent. As the effluent perks through the soil, it becomes harmless to the environment.

TOO MUCH WATER

Limit the amount of water entering the septic system to household sanitary drains. Don't discharge sump pumps or gutters into the tank. The increased flow of water disrupts the bacteria cycle. If an unusually high influx of water enters the tank, the septic field may begin to produce unpleasant odors. If you are using the septic tank only for household waste and the field begins to smell, inspect the tank. The tank may have a void allowing groundwater to enter it. This inspection will probably need to be done by a professional.

THE GREAT DISPOSER DEBATE

Garbage disposers raise considerable controversy when discussed in the same breath as septic systems. Some people say they will not harm the septic system. Some code enforcement offices ban the use of garbage disposers with septic systems. What is the truth when it comes to disposers and septic systems?

In general, it probably is not a good idea to use a disposer with a septic system. There are many factors to consider. Older systems were installed before disposers were popular and were not designed to handle them. Depending on the frequency of use, disposers can substantially increase the amount of solid waste entering the septic tank. This heavy increase could affect the function of the septic system. In the event disposer waste makes it to the drainfield without being properly broken up, it may clog the field, causing a system failure. I suggest you talk with septic system experts to determine a sound answer in your individual circumstances.

SEPTIC TANK ADDITIVES

Should you feed your septic tank additives? There may be some additives that will enhance your septic system, but I don't know what they are. My experience and research tend to indicate additives are not necessary and produce minimal results. Some additives are believed to create sludge bulking and interfere with the action of the system. If this happens, the system may suffer long-term effects from a short-term cure.

POTENTIAL PROBLEMS

Grease
Most of us know grease and drains don't mix well. The same is true of grease and septic systems. Bacteria is not very effective on grease, and the grease may cause blockages in the drainfield. Avoid putting grease in your drainage system with or without a septic tank.

Chemical Warfare
Chemicals may be your septic system's worst enemy. Strong chemicals are capable of causing serious trouble in the septic tank. The chemicals can prevent bacteria from decomposing the solids in the tank. If this happens, the entire septic system can become clogged with solids.

In-home photography darkrooms are a source of unwanted chemicals in the septic tank. If you have a darkroom, think twice before washing your chemicals down the drain. Repetitive use of chemical drain openers is another frequent cause of chemical corruption in the septic tank. Avoid letting chemicals into your septic system.

Not All Paper is Created Equal
Septic systems are designed to work with toilet paper. They are not meant to work with paper towels or other hard papers. If you are in the habit of flushing paper towels or similar products, stop doing it. The hard paper may not dissolve properly, causing potential for a clogged drainfield.

Poisonous Gas

Septic tanks hold methane gas. This gas is potentially explosive when an open flame is present. When the cover is first removed from a septic tank, the gas may be concentrated and volatile. Never remove the top of a septic tank with any type of flame or fire in the immediate vicinity. Methane gas is also hazardous to your health if breathed for any extended period or in high volume.

Inspections

Have your septic tank inspected regularly. Depending on the size of the tank and the use it sees, it should be inspected every one or two years. The best time to have the tank pumped is in the spring. Pumping the tank in cold weather can result in excessive buildup of solids. Bacteria are not as active in cold weather. The lack of bacterial activity will result in increased solid waste.

WHAT HAPPENS WHEN THE DRAINFIELD BECOMES CLOGGED?

If you fail to treat your septic system with respect, it may become clogged. When this happens, the effluent cannot go down into the soil. The only place it has to go is up, or back into your house. This can be a nasty problem.

When a drainfield is severely clogged, it must be replaced. This becomes a very expensive proposition. The gravel bed must be removed and replaced with new gravel. The slotted pipe may need to be removed and replaced. The disturbed area will have to be graded and seeded or sodded. All of this can add up to thousands of dollars.

As a system becomes old, replacement may be inevitable. Proper care of the system can prolong its life considerably. With the cost of replacing a drainfield, it should be worth your effort to take care of the one you have.

17

Valves, Sillcocks, and Miscellaneous Plumbing

This chapter deals with the small items in your plumbing system. If what you want to work on has not been covered yet, it is probably in this chapter. Unless otherwise noted, the water should be turned off to work on the following parts.

SMALL PLUMBING ITEMS

Boiler Drains

Boiler drains are used for a multitude of purposes. They can be purchased to screw in, solder on, or be fitted with compression nuts and sleeves. To disassemble a boiler drain, you loosen the packing nut and unscrew the stem. There will be a washer on the end of the stem. Replacing the washer or the packing is the only normal repair done to a boiler drain.

Gas Valves

Do not attempt to work on your own gas pipe and fittings. Call a professional when the work at hand has to do with gas.

Ball Valves

Ball valves are not normally used in residential plumbing. Ball valves do not have washers; they have a rotating ball that causes the flow to be open or closed. Ball valves give very little trouble and are replaced like any other shut-off valve.

Gate Valves

Gate valves are used at water heaters and on main water supply pipes. Gate valves do not have wash-

ers; they have a gate that moves up and down to open and close the valve (Figure 17-1). There is not normally any need to repair a gate valve. If the packing nut leaks, it can be loosened and repacked like the packing nuts discussed throughout the book. Replacement procedures are the same as for any shut-off valve.

Stop-and-Waste Valves

Stop-and-waste valves are used as cut-off valves in many instances. They are less expensive than gate valves and can be used to drain the pipe they control. Stop-and-waste valves do have washers (Figure 17-2). They also have a weep hole or bleed valve on the side. This weep hole is used to drain the pipe in front of the valve after the valve is closed.

There is an arrow imprinted on the side of stop-and-waste valves (Figure 17-3). The valve should be installed so that the arrow is pointing in the direction the water flows through the pipe. To replace the washer in these valves, the packing nut is removed and the stem unscrewed. The washer is on the end of the stem, held in place by a screw. The stem's packing may need to be replaced now and then. There is also a small rubber disk inside the weep hole cap. If the valve leaks from the weep hole cap, check to see if the rubber disk has gone bad or try tightening the cap.

Frostproof Sillcocks

Sillcocks are the fitting onto which you attach your garden hose on the exterior of your home. Frostproof

Figure 17-1.

Figure 17-2.

Figure 17-3.

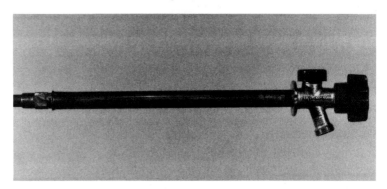

Figure 17-4.

Figure 17-5.

Figure 17-1. Gate valve. **Figure 17-2.** *Stop-and-waste valve with bleed valve.* **Figure 17-3.** *Stop-and-waste valve with arrow.* **Figure 17-4.** *Frostproof sillcock.* **Figure 17-5.** *Hose bibb.*

sillcocks are not supposed to freeze, but they sometimes do. When they freeze, it is usually because the sillcock is installed with inadequate drainage pitch. Frostproof sillcocks come in various lengths, ranging from 4 to 14 inches (Figure 17-4). The washer that closes the waterway of a frostproof sillcock is located at the back of the valve, near where the water pipe connects to the sillcock.

The principle behind a frostproof sillcock is simple. The valve is designed to cut off the water at a point where it will not freeze. When the water is cut off, excess water is meant to drain out of the valve opening. If the sillcock is installed without the proper drainage pitch, the water will not drain out. When the sillcock is level or pitching back toward the home, the water cannot drain. Water left standing in the body of the sillcock can freeze.

The length of the sillcock is determined by the thickness of the wall in which it is mounted. If you have a 6-inch wall, the sillcock should be at least 6 inches long. Ideally, the sillcock should be 8 inches long. This allows the cut-off portion to extend through the finished interior wall and into a heated space. If the sillcock terminates within the exterior wall, it and the pipe supplying it are subject to freezing.

Changing the washer in a frostproof sillcock. The washer for a frostproof sillcock is attached to the bottom of the stem like most other valves. The stem removal is accomplished by the same methods used on other valves. The difference is the length of the stem. The sillcock has a stem that runs the length of the sillcock's body. In some cases, the stem can be over a foot long. The stem is removed by loosening the packing nut and unscrewing the stem.

Replacing a frostproof sillcock. If a sillcock needs to be replaced, you may have to open the wall. When the back of the sillcock protrudes into the home, you can make the swap without cutting the wall. Most sillcocks are soldered onto copper pipe or attached with screw-type fittings. If you cannot see the connection between the water supply and the sillcock, you will have to open the wall to expose it.

Once you can see the connection, replacement is not difficult. As always, cut off the water to the sillcock. Open the sillcock valve and allow any excess water to drain. Remove the screws that anchor the sillcock to the home. Either unscrew the sillcock from its connection or sweat it off the copper. Install the new sillcock by reversing the removal process.

Hose Bibbs and Regular Sillcocks

Hose bibbs, also known as regular sillcocks, are very similar to standard valves (Figure 17-5). They do not have extended stems and are not meant to resist freezing. They use either screw-type connections or soldered connections. Faucet washer replacement is done by loosening the packing nut and unscrewing the stem.

ROUTINE PLUMBING PROCEDURES

By now you should be developing a strong sense of how plumbing works. There are countless variations of similar products in the plumbing field, but most of them work on the same principle. When you grasp the basics of plumbing, there will not be much you cannot figure out on strange items. If you don't have detailed instructions from an owner's manual, there are times when you should call in a professional. Sometimes the trial-and-error method of repair results in a costly lesson.

If you have understood this book, you should be able to accomplish many small plumbing jobs. The information given can be applied to a wide range of plumbing. For example, most items turn counterclockwise to loosen and clockwise to tighten. In most cases, the water supply should be cut off to the device you are working on. A great deal of plumbing is common sense.

The remainder of this book deals with planning to remodel your plumbing, designing new plumbing, evaluating professional plumbers, and practicing routine maintenance. This chapter concludes the hands-on, how-to-fix-it work. What follows is advice on using your head to avoid trouble.

18
Adding to Existing Plumbing Systems

When you remodel or add space to your home, you may add plumbing fixtures. These fixtures can put a strain on the existing plumbing system. In some cases, your sewer pipe may not be large enough to handle the demands of an additional bathroom. The septic system may not be capable of supporting the addition of new plumbing. Before you increase the number of plumbing fixtures in your home, you should consider the size and condition of your existing plumbing.

REPLACING A BATHTUB WITH A SHOWER

On the surface, replacing a bathtub with a shower may not seem like a strain on your plumbing. After all, both are bathing units, what difference could it make? Well, the difference is in the size of the drain required for each of these fixtures. Bathtubs can use a drain with a 1½-inch diameter. Showers require a drain with a 2-inch diameter.

This seemingly subtle difference can balloon into significant work. If you connect the shower to an undersized drain, the shower may flood your bathroom. When you are changing the type of plumbing fixtures currently in use, code enforcement inspections are usually required. Hooking the shower up to a 1½-inch drain will fail the inspection in most jurisdictions. This failure will result in the need to run new 2-inch pipe from the shower to the drain it connects to. Depending on location and accessibility, this can be quite a job.

Another difference between tubs and showers is the location of their drain. Most tubs have their drain located at the end of the tub where the faucets are. Most showers have their drain in the center of the shower base. Obviously, the drainage piping will have to be run to a new location to allow for the installation of a shower. Generally, this is not a problem, but it can be. Suppose your new shower is set so that its drain is positioned over a floor joist or light fixture. There must be adequate room for a 2-inch trap to connect to the shower's drain. All these little particulars can add up to a costly conversion. This type of change-over is only one example of how remodeling or adding to your plumbing can get complicated fast.

SEPTIC SYSTEM CONCERNS

In Chapter 5, I talked about garbage disposers. When kitchens are remodeled, they are customarily updated. Garbage disposers are one of the more common add-ons found in remodeled kitchens. As you learned earlier, there is a controversy over the effect a garbage disposer can have on a septic system. Another consideration when adding a garbage disposer is the type and condition of the drainage piping.

When the kitchen drain is old galvanized pipe, a garbage disposer can cause frequent stoppages in the drain. This type of problem is avoidable if you look ahead. If the kitchen is piped with galvanized pipe, replace the drainage piping when remodeling.

While the remodeling is in progress, replacing the drain is not a major job. Once you have a new kitchen, tearing into it to replace the drains can cause severe stress.

When septic systems are designed, their design is based on the intended use as given to the designer. If you choose to add an additional bathroom, the requirements of this extra plumbing may not have been calculated in the original design. Increasing the demands on the septic system may cause it to fail. If your septic system fails, you could be looking at several thousand dollars to repair or replace it.

Before you spend your money to add new plumbing, have the septic system inspected by professionals. Obtain a report on the capabilities of the existing system. If you spend a lot of money for a new bathroom only to have to spend just as much to upgrade the septic system, you could become depressed very quickly. Some preliminary research and planning can limit your exposure to this type of aggravation.

The Building Drain and Sewer

Plumbing codes require specific pipe sizes to be installed to handle the load of a plumbing system. The proper pipe size is determined by the number of fixture units on the system. There are charts in the code books that enable professional plumbers to size a plumbing system.

Some homes have a building drain, and/or sewer, with a 3-inch diameter. By most codes, this is fine for up to a two-bathroom house. Adding a new toilet to a 3-inch system will cause considerable cost. By average regulations, 3-inch pipe cannot receive the discharge of more than two toilets. Adding a third toilet will require the replacement of the 3-inch pipe with a 4-inch pipe. The change will have to be made at the point of connection for the toilet closest to the sewer.

If your main sewer is a 3-inch pipe, the conversion requires digging it up from the house to its destination. This type of work gets very expensive very quickly. Again, something as simple as the size of your sewer may not loom large in your plans, but it

can crunch your budget when the code officer pays you a visit.

All these considerations are ones that a professional plumbing contractor should explain to you. When you plan your plumbing remodeling, your plumbing contractor can be of great help. The local code enforcement office will also be willing to assist you. As the homeowner, you are the one spending the money and suffering the consequences of poor decisions. You should make a point of investigating all avenues of your plumbing remodeling before making financial commitments.

ADDING A BASEMENT BATHROOM

Basement bathrooms are a frequent cause of do-it-yourself nightmares. These installations often incorporate the use of a sewage-ejector pump. The basin these pumps sit in must be vented. The vent is usually a 2-inch pipe that extends above the roof or ties into an existing vent in the attic.

On paper, this vent does not seem intimidating, but it can cause plenty of problems. The vent must go up from the basement through the house until it reaches a tie-in point or the roof. For the unsuspecting homeowner, the installation of this vent can spell destruction to the upper levels of the home.

Walls may have to be opened to allow the pipe to pass through them. Without strong remodeling experience, even professional plumbers can make a mess of your home installing these vents. Trying to do it yourself can be tedious, dangerous, and costly. There are electrical wires hidden in most walls. If you cut into one of these wires, you may not live to worry about the vent installation. The vent is only part of the complexity of a basement bath.

The sewage-ejector pump requires a basin in the floor. This means your concrete slab will have to be broken up and the dirt excavated. Again, on paper this is not overwhelming, but it can turn bad. I have opened up basement floors to find fast-flowing streams below them. The water moving under the floor has been so severe, we were forced to use cast iron pipe. There was no way to glue plastic pipe together in the water. The pump basin was floating up out of the hole, and the water's force was enough to move gravel and dirt as it passed.

This particular job not only presented special problems, it caused me to lose a lot of money. I had never anticipated finding a stream under the basement floor. Everything about that job became expensive. As a professional, I was able to complete the installation and repair the floor, but it was not easy or cheap. It takes a little thought to figure out how to repair a concrete floor when there is water running through it. Admittedly, this was a unique situation, but strange things are known to happen in remodeling endeavors.

ADDING NEW DEMANDS TO WATER DISTRIBUTION PIPES

The water distribution system in your home normally consists of $7/8$-inch and $5/8$-inch pipe. These pipes are known in the trade as $3/4$-inch and $1/2$-inch pipe. Half-inch pipes generally are only allowed to feed two fixtures. This code regulation can cause the remodeling homeowner sizable expense.

When professional plumbers install a new plumbing system, they typically make the installation to meet *minimum* code requirements. This saves money and makes their estimates to the general contractor appealing. It is not until you wish to add on to the existing plumbing that their bargain price is no longer a good deal.

Consider this example. You are about to install a new bathroom in your attic. The carpenters have added dormers, and your old attic is about to become two additional bedrooms and a new bathroom. When you planned your plumbing, you planned to connect into the plumbing at the bathroom below the conversion. Under standard conditions, you probably never considered the sizing of the existing water distribution system.

When the time comes to install the water piping for the new bathroom, you are shocked. Your plumber presents you with an estimate that is well over your budgeted figures. When you question the plumber,

you find the reason for the big price difference. You planned on making connections to the plumbing almost directly under the new bath. To comply with code requirements, the plumber must make connections where $3/4$-inch pipe can be found, and this is some distance away.

Three-quarter-inch pipe costs more than $1/2$-inch pipe, and the distance to a legal connection point is much longer. The labor required to run the pipe is expensive, and the total estimated plumbing cost is formidable. You have very few options. Adding two new bedrooms without a bathroom is not very practical. Adding the bathroom is going to cost much more than you thought. This problem might have had a different outcome if it had been addressed in the planning stage.

When you are planning to add new plumbing, you must be aware of the condition and size of your existing plumbing. As with all the code information in this book, your local code officer may be working from a different set of rules. Plumbing codes are written for specific locations, but local code officers have the freedom of interpretation. The interpretation may have a strong influence on what is and is not acceptable in your area. To be safe, always consult professional plumbers and the local code officer before moving ahead with your plans.

WATER HEATER DEFICIENCIES

When your water heater was installed, it was sized for the plumbing being installed at that time. If you add new plumbing, you may have to make adjustments for heating your water. For homes with electric water heaters, the water heater capacity is typically between 40 and 52 gallons. Adding a whirlpool or a new bathroom may place too much demand on the existing water heater if the use of hot water will grow.

If the number of people using hot water does not increase, and the frequency of use stays the same, adding a bathroom does not make increased water-heating capacity compulsory. The size of a water heater is directly related to the amount of use required. If a bathroom is added for the convenience of the location, not to accommodate heavier plumbing usage, there is no need to upgrade the water heater.

A whirlpool tub can definitely put a strain on a standard water heater. Whirlpool tubs can hold more water than a 52-gallon heater can heat. This type of remodeling requires advance planning if you are to be happy with the results.

Connecting to Existing Plumbing

Adding new plumbing can present challenges in connecting it to the existing plumbing. You will need access to pipes of an approved size to connect the new plumbing. This can require opening walls, ceilings, and concrete floors. Before you get too enthralled with adding new plumbing, make sure it is feasible. Consulting with professional plumbers who are experienced with remodeling is one of the best ways to do this.

IN SUMMARY

Whenever you are involved in remodeling, unexpected events can occur. If you are not careful, the expense of these learning experiences can be overwhelming. To avoid sudden surprises, approach your remodeling efforts with extensive planning. Remodeling work is very profitable for experienced plumbers. These plumbers know how to evaluate the requirements of the job and can make a difficult task easy.

If you attempt to perform your own remodeling plumbing, you may find the results less than desirable. There are a number of potential pitfalls and frustrations. Extensive plumbing remodeling should be left to the professionals. These contractors, if they are seasoned, have already paid the price for their expertise. As a homeowner, you should seriously consider hiring professional plumbers to design and install your plumbing for remodeling projects.

What looks like a simple task in the beginning can put a real damper on your overall job. I have been involved with plumbing and remodeling for over

seventeen years, and I still come across situations I have never encountered before. With this much experience, you would think I have it down to a science, but there are no hard-and-fast rules in the remodeling of plumbing. Every job offers its own set of circumstances. With experience, I know what to expect in most cases and how to overcome the problems. As a weekend plumber, you could be at a total loss and might do more harm than good.

For example, I have sawn open walls to install plumbing and come face to face with hundreds of angry bees. I have gone into attics to run vent pipes and been swarmed by bats. When crawling around under houses, I have encountered rattlesnakes, skunks, and feral cats. These situations are not in the normal game plan. I have also seen existing plumbing that was not worth connecting to. These systems were illegal and unhealthy. There are far too many variables for you to expect to be a competent remodeling plumber without training.

This is not to say that you cannot perform the duties required to add to your plumbing system. It does mean that there is much more involved with the remodeling of plumbing than meets the eye. If you elect to do your own work, plan the work carefully. If what you install has to be ripped out because of code violations, how much money did you save? Coming eyeball to eyeball with an aggressive snake in your crawlspace can quickly make you hang up your torch. These are only some of the trials and tribulations you will work with in adding to your system.

I advise you at least to consult with professional plumbers and the code enforcement officer before attempting to add to your own plumbing. There is no satisfaction in failing. With enough preliminary planning, there is no reason your remodeling efforts should fail. The odds of successful remodeling are greatly increased when you use experienced remodeling professionals. Doing the work yourself leaves you open to a number of possible problems, but it can be done.

The final decision is yours. I will tell you, I cannot count the number of times I have been called in to correct the best efforts of well-meaning weekend plumbers. Not all these jobs were done by home-owners. Many of them were done by inexperienced part-time plumbers. While the plumbers may have been well-qualified to plumb a new house, they did not know how to remodel existing plumbing.

Plumbing for remodeling jobs requires a special talent. It is mixed with patience, experience, and creative techniques. These qualities cannot be taught in a book; they must be learned on the job.

Choosing the right plumber for a remodeling job is not as simple as turning to the Yellow Pages. The plumber must meet certain criteria, and this information is described in Chapter 20. Unless you are extremely skilled, I recommend calling in a proven professional for your remodeling efforts.

19
Designing a New Plumbing System

When you plan to add substantial new plumbing or replace an outdated system, the design of the new plumbing is significant. The proper design will work better and cost less than a poorly designed system. While the technical aspect of engineering a plumbing system requires special skills, there is much the average person can do to enhance the layout of his or her plumbing system.

When most people think of replacing an old system, they only intend to put new plumbing where the old plumbing is. This may be the best way to do it, but don't count on it. If you are going to the expense of replacing a complete plumbing system, you might as well improve more than the quality of the pipe and fixtures.

The location of plumbing fixtures can make a big difference in the convenience and serviceability of a plumbing design. The route of the pipes affects the cost and function of the system. The size of the pipes plays a large role in how satisfied you are with the new plumbing. Since new fixtures are introduced each year, you may find a style that suits your situation better than existing fixtures do. With so many variables, planning a new plumbing system deserves special attention.

FIXTURE STYLES

The style of your new plumbing fixtures can make your life easier and more enjoyable. When you begin to plan your new system, there will be a wide variety of fixtures from which to choose. Some of the differences are technical, some subtle, and some variations are quite noticeable. You should anticipate spending time shopping for the fixtures that will work best for you.

Toilet Styles

Most homes use a toilet that sits on the floor. Wall-hung toilets are not common in modern residential applications. This is about the only thing you can count on to be standard in the style of a residential water closet. You can choose a bowl with a round front or an elongated bowl. Round-front toilets are used more often, but elongated toilets may be more pleasing to you. The seat on the toilet may be plastic, wood, or a cushioned material. The seat is strictly a matter of personal preference.

The obvious difference in toilets is their aesthetic appeal. A technical difference is the amount of water required to complete a satisfactory flush. If you have back problems, a toilet that sits high above the floor will be more comfortable for you. If you despise cleaning between the tank and the bowl, a one-piece toilet can ease your cleaning chores. All these options can cloud your choice. Product manufacturers produce literature you can study to understand the functional differences in their products. Plumbing suppliers often have assorted fixtures on display. Review the technical information and make the rounds to see the products available.

Before deciding on a specific product, talk with your plumber. Plumbers work with different plumbing fixtures all the time. Your plumber may be able to help you in the final decision on which product to buy. If a particular style requires an unusual amount of service, a professional plumber will know it. If the plumber is called regularly to repair a particular brand, you may want to consider a different brand. As a professional, your plumber can give you advice on which items work best under given conditions.

Lavatory Styles

Do you want a pedestal lavatory or a vanity with the lavatory molded into the top? Will a self-rimming lavatory be more enjoyable than a rimmed lavatory? Does the depth of the lavatory bowl make a difference to you? When you plan your lavatory purchase, there will be many choices.

The deciding factor in the lavatory is normally one of taste. The decision may be influenced by the amount of space you have to work with. Follow the same shopping advice given for toilets to decide on your lavatory.

Faucet Selection

The type of faucet you choose can hinge on several factors. Appearance is a key reason for choosing one faucet over another. The ease of repairing the faucet should be a consideration. Your preference for a single handle or multiple handles will also affect your decision. The product's warranty may swing your decision. Consulting with your plumber may reveal facts on which to base your decision.

In some cases, safety may have a bearing on your faucet selection. If you have young children or elderly persons using the faucet, an anti-scald faucet is a good idea. These temperature-controlled faucets protect the user from being badly burned with hot water. If you have arthritis, the design of the faucet handles can make a huge difference in your comfort. Read the brochures and look through the showrooms to gain a breadth of knowledge.

Bathing Units

Bathtubs and showers see frequent use. Their design can have a direct impact on your comfort and safety. When considering your options for bathtubs and showers, look below the surface. Investigate the materials used to make the bathing unit. Decide if the shower base or the bottom of the tub will be more slippery than its competitors. Do you want your new shower to have a seat in it? Will you be able to get the unit into the new bathroom? Trying to get a one-piece tub-shower combination into an existing home is all but impossible.

You may have to look to a sectional unit when working with existing conditions. If a sectional unit is planned, determine how the sections go together. Deliberate the risk of leaks with each type of unit inspected. If you are looking at tub surrounds intended to be glued over existing walls, look for quality. Cheap wall surrounds cause much more trouble than the money you save is worth.

When debating on the type of materials you want the tub or shower made from, consider location. On a second floor, a steel bathtub can make quite a racket when the shower head is running. This pinging can be heard in the room below the tub. Perhaps you will want to replace your old tub with a whirlpool tub. All these alternatives are available; it is up to you to decide what you want and can afford. With extensive research and planning, you can make a wise decision.

Kitchen Sink Selections

The right kitchen sink can make your kitchen duties more fun. Kitchen sinks are available in a multitude of designs. You can have one bowl or two. You might opt for a special bowl, just for your garbage disposer. Maybe you would enjoy a sink with a custom cutting board or extra depth. Stainless steel is the typical choice for the sink's finish, but you certainly are not limited to it. Cast iron sinks are available in a vast range of enameled colors. There are sinks available for every conceivable need and desire.

Pipe Sizing

On a major installation, pipe sizing can be confusing. If you are using a professional plumber, he will size the piping for you. If you attempt to do the work yourself, refer to the charts in Appendix 3. The charts give you an idea of what your piping needs will be. Since plumbing codes differ, don't take these charts literally. Check with the code enforcement office before installing the new plumbing.

The job of replacing or installing a complete plumbing system usually requires the services of a professional. While you may be able to obtain the necessary permits to do the work, installing the plumbing to comply with the plumbing code is not easy for the average person. There are hundreds of rules and regulations affecting the installation of plumbing. I strongly recommend the use of professionals on a job of this magnitude.

Pipe Type

In almost every case, the drainage pipe will be a schedule 40 plastic pipe. Cast iron is sometimes used to control noise in homes with multiple levels. This practice is uncommon but is used in some custom homes. The water distribution pipes might be any of the approved materials discussed earlier in the book. My choice would be either copper or polybutylene.

Pipe Routes

If you employ the services of a professional plumber, he will lay out the piping. If you attempt to do the job yourself, you will have to understand the local plumbing code before you can map the pipe locations. The technical aspect of routing the pipes is too complicated to approach in this book. Without the knowledge of what code you will be working with, I have no way to help you design the piping.

Selecting Fixture Locations

The location of your plumbing can have a strong effect on your happiness with the fixtures. Installing your laundry facilities in the basement has advantages and disadvantages. The advantage is that you don't have the appliances in your primary living space. The heat from the dryer will not affect your home as much. Noise from the appliances will not be as noticeable. If the washing machine leaks, expensive flooring will not be ruined. The disadvantage is that you have to carry the laundry up and down the basement steps. This creates a safety risk and can become tiring. Before blindly picking a spot for your laundry hookup, weigh all the considerations.

Fixture Noise

Installing a bathroom over another room can produce annoying noise. When the toilet is flushed, you will probably hear it. If the drain runs down a wall near the lower room, you might hear the water passing through the pipe. As the shower is being used, you may have to listen to the constant pinging of water on the base of the shower. These noises can become aggravating at certain times. Do you want to be entertaining and hear the toilet every time it is flushed? When you are designing your new plumbing, keep noise factors in mind.

Fixture Accessibility

During the initial installation, accessibility may not cross your mind. During installation, the walls are usually open and space is abundant. This all changes when the job is finished and there is a problem with the concealed plumbing. Look ahead and plan access routes for plumbing repairs. Make notes of how the plumbing was installed. These notes will make troubleshooting and repairing the plumbing later much easier.

Making the Most of Your Space

If you have limited space to work with, look to specialty fixtures. By placing the toilet or the shower in a corner, a small bathroom can be much more comfortable. Pedestal lavatories can eliminate the need for a massive vanity cabinet with sharp corners.

YOUR ROLE IN DESIGNING THE PLUMBING SYSTEM

If you give it enough thought, you can be very helpful in designing your new plumbing. While the technical aspects are best left to professionals, there is much you can do to help. Plumbers are sometimes consumed by code compliance and technical designs. They don't always think of the little items that can mean so much.

Your plumber may not think to ask if you would like a tall toilet to ease the strain on your back or legs. He probably will not give a second thought to putting the shower head at the standard height. If you are unusually tall or short, the location of the shower head may greatly affect your satisfaction with the final results. Telling the plumber to put the shower head lower before the work is installed does not cause a problem. Asking to have it lowered after the job is done can be quite another story.

It is your job to choose the fixtures you like. As the homeowner, you must tell the plumber about any special needs or desires you have. Plumbers do their job by the book. The book I am referring to is actually two books — the code book and the rough-in book. Code books outline the rules and regulations for plumbing. Rough-in books tell the plumber where a shower head should be placed. The code book is somewhat inflexible, but the guides in a rough-in book can be altered.

It is usually simple things that complicate a plumbing job. A plumber will install an outside faucet where he thinks best. The homeowner wants the sillcock closer to the garage or beside the deck. Either of these locations meets code requirements, but the sillcock should be where you, the customer, want it. Preliminary thought and open communication are the keys to obtaining a new plumbing system you are happy with. If either of these steps is omitted, the results can be less than desirable.

20

How to Choose a Professional Plumber

At some point during the average homeowner's life, the need for a professional plumber arises. The timing for this need is rarely good, and the pressure to find a plumber fast may be a factor. If you have a water pipe flooding your home and the cut-off valve won't work, time is of the essence. At these times, most people consult the Yellow Pages in their phone book. They run through the ads, calling all the plumbers, until they find one who is readily available.

Under emergency conditions, there is no time to pick and choose your plumbing contractor. The first available plumber is the one most often chosen. While this is understandable, it is not the best way to choose a plumber. Plumbers can be like any other professionals. They frequently specialize in specific aspects of plumbing.

A plumber who concentrates his efforts on installing new plumbing in commercial buildings may not want to come over and fix your toilet. A well-pump specialist may have no idea where to begin when asked to remodel the plumbing in your home. The point is, not all plumbers are the same. In densely populated regions, specialization is common. In remote rural areas, specialization is rare. In country settings, your plumber may also be your carpenter and electrician.

Before you can choose the best plumber for the job, you must determine what the job entails. Plumbers typically divide their services into standard catego-

ries. It is important to match the right plumber to the right job.

COMMERCIAL PLUMBERS

Commercial plumbers almost never do residential work. It is not that they couldn't, they just don't. These plumbers specialize in large plumbing jobs. They work with plumbing for hospitals, schools, shopping malls, and other types of commercial-grade plumbing. These plumbers are generally not a good choice for household repairs and installations.

REMODELING SPECIALISTS

Plumbers who specialize in residential remodeling are the best choice when you are remodeling. They know all the tricks to save you time and money. They can work magic with existing plumbing and can install new plumbing with less damage to your home than plumbers with less remodeling experience. These plumbers are not cheap. Their special skills provide them with a reputation that keeps them busy throughout the year.

While these plumbers are quite capable of doing repair work and new construction work, they are usually too busy with remodeling jobs to accept other assignments. Because of their specialized expertise, they can make much more money in remodeling. It is not that their price is significantly higher than another type of plumber, it is that their skills allow them to do the job faster. Doing contract work fast is how big money is made.

SERVICE PLUMBERS

Service plumbers save the day when you have an emergency. These are the plumbers who fix the plumbing you can't. They are your best choice for routine repairs and some replacements. They spend their lives fixing a variety of plumbing problems, so they are well-trained and fast at their work. Since these plumbers charge by the hour, their speed saves you money.

Asking a new construction plumber to troubleshoot and repair your old faucet may be asking too much. New construction plumbers install plumbing but rarely have to repair it. By the time they figure out what is wrong with the old faucet, or how to get it apart, a service plumber might have finished the job. When all you do is repair existing plumbing, you gain product knowledge and learn shortcuts.

NEW-CONSTRUCTION PLUMBERS

These plumbers make a living by installing new plumbing in homes being built. They are aware of plumbing code requirements for installations and are skilled in their field. If you are building a new house or an addition, these plumbers are a good choice. If your job requires extensive work on old plumbing or getting the new plumbing up through the existing house, remodeling plumbers will have an edge on new construction plumbers.

New construction specialists are generally not a good choice for repair work. Their life is spent installing new pipe and fixtures. They are not equipped to work with old parts. Their trucks are not stocked with all the washers, screws, and fittings needed to perform elaborate service work.

WELL PUMP PROFESSIONALS

Well pumps require special knowledge to troubleshoot. There are many electrical aspects to the pump that affect its performance. The plumbers who specialize in pumps can really shine when compared to a plumber who is not accustomed to troubleshooting pumps. Some plumbers don't have the proper meters and gauges to rate a pump's performance, or lack of it, effectively.

As in any job, the proper tools make the job go faster. Since you are paying an hourly fee for these plumbers, time is money. A poor diagnosis could cause you to buy a new pump when it is not necessary. Finding a plumber with pump experience can save you hundreds of dollars.

WATER-CONDITIONING MECHANICS

Working with water conditioners is another aspect of plumbing in which some plumbers specialize. Water conditioners are sensitive pieces of equipment. They have a lot of controls and settings that affect their efficiency. If the person working on your equipment is not well-trained, the equipment can be badly damaged.

There are companies that do nothing but treat water. While the average plumber may be able to work on your equipment, a specialist is a sure bet.

GENERAL PRACTITIONERS

Don't think that you *must* find a specialist. Plumbers with enough experience can perform any of these duties. In many cases, there will not be a specialist available. Small towns cannot support enough plumbers to allow for specialization. The key is finding a plumber with strong experience in the area with which you need help.

WHERE TO LOOK FOR YOUR PLUMBER

The phone book is an obvious place to look when searching for a plumber. Community bulletin boards are sometimes filled with cards and advertisements placed by plumbers. The local newspaper is another source of information for finding a plumber. All these sources will provide you with names and phone numbers, but they will not tell you much about the plumber.

The advertising may tell you what type of work the plumber is looking for. It may tell you if he offers twenty-four-hour service. What the advertisements won't tell you is whether the plumber is dependable, fair, and professional. Talking with your friends and neighbors is an excellent way to locate a plumber for your job. Your acquaintances may have had experience with a plumber. They can help you to evaluate the plumber's credentials before you spend your money.

When you are planning to spend large sums of money, it helps to know what you are spending it for. There is nothing worse than paying a plumber big money only to be disappointed. By talking with your friends and associates, you can limit your risk. There is no guarantee their plumber will treat you the way he did them, but the odds are better you will know what you are paying for.

REFERENCES

Most people never take the time to check references, but they should. It is true that what you are told may not reflect the service you get, but it is one more step in reducing your risks. Any established plumber should have numerous references available upon request. Be sure the work done for the references was similar to work you will be having done. Just because a plumber did a great job replacing a toilet doesn't mean he is the best plumber for your remodeling efforts.

WORKING WITH NEW BUSINESSES

There are pros and cons to be considered when working with a young business. A plumber just starting in business should be aggressive for your business and your satisfaction. When first starting out, it is imperative to please the customer. Customer satisfaction should remain a priority for all businesses, but as companies prosper, they sometimes forget who made it possible for them to develop their business.

New companies will frequently work for less money than established firms. If you do your homework and monitor your business relationship, an up-and-coming business can present excellent value. That's the good news. The bad news is, new companies are considered a higher risk than established firms.

The risk of a new business failing is greater than that of a proven business. The business ethics and reputation of older companies are easier to judge. If you use contracts for large jobs and don't give substantial deposits for the work to be done, you can usually do all right with a new company.

What you are looking for is the experience of the person doing your work. If the individual is an excellent plumber, it doesn't matter if he has owned his own business for six months or six years. The only risk you run that is uncontrollable is the risk of warranty work. If the company goes out of business, they will not be there to provide free warranty service. All the other risks can be removed with good contracts and cash control.

WORKING WITH PART-TIME PLUMBERS

Moonlighters can offer extremely low prices for their work. They typically earn their living at a job and do plumbing in their off hours for extra money. The quality of a moonlighter's work may be every bit as good as that of a full-time plumber. While you can save money, the risk of having trouble with the plumber escalates.

What happens if the part-time plumber's work springs a leak when he isn't available? Will the part-timer be available for warranty work? These are only two of the possible pitfalls of working with part-timers. I would not discourage you from working with a part-time plumber, but I do advise you to be careful.

QUESTIONS TO ASK ALL PLUMBERS

Licenses

Is the plumber licensed for the work you want done? If the plumber is unlicensed, avoid using him.

Plumbers are required to be licensed in most places, and you should only allow licensed plumbers to work for you. What type of license does the plumber hold? A plumber with an apprenticeship license is required to work under the direct supervision of a journeyman or master plumber. Journeyman plumbers are allowed to work without direct supervision, but they must be supervised by a master plumber. A master plumber can work without any supervision; he is at the top of the ladder.

Insurance

Is the plumber insured, and what does the insurance cover? All plumbers should have a liability insurance policy. If they have employees, they may need worker's compensation insurance. Verify the status and coverage of the plumber's insurance to protect yourself and your assets.

Bonding

Is the plumber bonded? Bonding is a form of insurance to protect you. Inquire about the status of the plumber's bonding. Bonding may not be required by your state, but it is worth asking about.

Warranties

What assurances will you have that the plumber will stand behind his work? Many states have laws pertaining to warranty work. The time periods and specifics of the laws vary.

Mechanic's Liens

If the plumber is not paid for work done on your home, he has the right to place a lien against the property. Always have the plumbing contractor sign a lien waiver each time he is paid on a large job. Lien waivers are a form that a contractor signs to acknowledge full payment for all labor and materials provided for a job. Your attorney can provide you with a lien waiver form.

Materialman's Liens

If the plumber does not pay his supplier for materials installed in your home, the supplier may have the right to lien your property. Ask the plumber to provide proof of payment for all materials installed in your home.

Emergency Calls

Ask your plumber how he can be reached in case of a plumbing emergency. If you have to go through an answering service and wait until the next morning, look for another plumber. Any reputable plumber should be willing to respond to an emergency call immediately if the emergency is the result of his work. If the emergency is unrelated to work installed by the plumber, the plumber may not offer twenty-four-hour service.

IN CLOSING

The quest for the perfect plumber can be an arduous one. Plumbers are known for their independent nature. They can be difficult to understand at times and even more difficult to find. Excellent plumbers stay busy. Shop early so you can give the plumber plenty of notice before your work is needed. Be cautious when giving advance deposits and require written agreements for all extensive work.

21
Avoid Costly Repairs with Routine Maintenance

Plumbing will last for many years when it is not abused. To keep all your plumbing in top condition, follow the directions in your owner's manuals for the plumbing products. Most manufacturers provide detailed maintenance instructions with their products. For items without manuals, use common sense and the advice from this book.

Much of your home's plumbing does not require any regular maintenance. The parts of your plumbing that are most likely to need attention are noted in this chapter. The maintenance advice given here is general advice. Always refer to the owner's manual for your particular products before servicing them. Some plumbing is very sensitive and requires unique methods to maintain. Never use this, or any other generic book, to override the instructions given by a product's manufacturer.

DRAIN, WASTE, AND VENT PIPES

These pipes do not require any type of routine care. Unless they are giving you a problem, they can be left alone for years. It is wise to inspect the piping periodically. Look for leaks, sags, or other unusual conditions. Avoid using chemical drain cleaners in your drain and waste pipes. The chemicals in the drain opener may have an adverse effect on the piping.

WATER-DISTRIBUTION PIPES

Under average conditions, these pipes do not require any regular attention. It is a good idea to

inspect them from time to time, but there is rarely anything to be done with them. If you have metallic pipe in your water-distribution system, it could pay to have the water tested. If your water contains extreme amounts of acid or minerals, you might want to invest in a water conditioner. Over time, poor water quality can deteriorate your water pipes. In general, there is very little to do in the maintenance of water pipes.

SEPTIC TANKS

The care of septic tanks is described in Chapter 16. Avoid putting too much strain on the septic system by adding additional plumbing. Have your tank inspected annually to determine when it should be pumped. Do not allow strong chemicals to go down the drains and into the tank. Be careful not to dispose of items in the drains that will clog the septic system. If you observe the advice in Chapter 16, you should not encounter serious problems.

BACKFLOW PREVENTERS AND PRESSURE-REDUCING VALVES

These devices do not require any routine maintenance.

ELECTRIC WATER HEATERS

An electric water heater should be partially drained about once a month. By draining some of the water from the heater, mineral deposits are removed. This extends the life of the tank and the lower heating element. When water heaters are not drained on a regular basis, rust and mineral deposits build up. These deposits can build to a point where they block the drain and encrust the heating elements.

It is a good idea to turn off the power to the water heater before draining it. When the power is left on, there is some risk that the elements may be burned out by draining the tank too far. If the elements do not remain submerged in water while the power is on, the elements will be ruined.

IN THE KITCHEN AND BATHROOM

The basic plumbing for a kitchen does not require any attention. The kitchen faucet may need to be attended to if it develops leaks, but the plumbing is fairly self-sufficient. There shouldn't be any kitchen plumbing that requires you to perform predictable maintenance.

Bathroom plumbing is similar to kitchen plumbing in regards to maintenance. There is no need for any type of routine maintenance.

TIPS AND POINTERS

While very little of your plumbing requires regular maintenance, there are ways to extend the life of your plumbing through respect for the plumbing.

Faucets

If a faucet begins to drip, it should be repaired. If you put off the repair and force the handles to tighten them in order to retard the leak, you can damage the faucet stem and seat. Putting off the repair may mean a more extensive rebuilding when you do get around to the repair.

Allowing water to drip around the faucet's spout can cause water damage to the cabinet holding the sink and to your floor. If this damage goes unchecked, the carpentry work required to repair the damaged wood can be expensive. Leaks of this nature can also cause the faucet's mounting nuts to rust. When this happens, removal of the faucet becomes quite

difficult. You may have to cut the mounting nuts off with a hacksaw.

Cracked Caulking

If during your routine inspections you notice cracked or missing caulking, replace it. Caulking prevents water from going where it shouldn't. When the caulking is defective, renegade water can cause heavy and hidden damage. This is especially true around bathtubs and showers.

Tile Grouting

When your tub or shower has tile around it, the grouting between the tile may act as a sponge. The grouting is the substance placed between tiles to block the passage of water. With age, grouting loses its ability to prevent water from passing between the tiles. This fact is true for wall tile and floor tile.

When you step out of the bathing unit, water drips from your body. If the grouting is bad on the floor tiles, this water seeps past the tile and into the wood subflooring. Ultimately, the water may appear in the ceiling of the room below the bathroom. With wall tile, water from the shower head bombards the tile. When the grouting is weak or missing, the water invades the wall. All this water can do serious damage to the structure of your home.

If you have tile in your bathroom, inspect it as you clean it. Look for areas where the grouting is missing or cracked. If you notice water soaking into the tile floor, have the grouting repaired. This is not a plumber's job, but it is important to maintain the structural integrity of your home.

Toilets

As time passes, toilets sometimes develop leaks at the wax seal. This leak can go unnoticed until the floor under the toilet is rotted. During your cleaning, inspect the flooring around the toilet. If you have reason to believe the toilet is leaking, test for the leak. Place a ring of toilet paper around the toilet's base. Flush the toilet several times. If the toilet paper around the toilet becomes wet, replace the wax seal. When a rotted floor is suspected, you can probe the floor with the point of a knife. If the knife sinks quickly into the subfloor, it is a strong indication that you have water damage.

To prevent stopped-up toilets, be careful what is flushed down the toilet. Do not reduce the amount of water used in the flushing too dramatically. Many times, people reduce the amount of water in the toilet tank to conserve water. Without adequate water, the toilet cannot function properly. Even if the toilet appears to flush well, the solids may become lodged in the drain. Maintain the recommended water level in your toilet's tank.

Lavatories

Shampoo bottle caps and toothpaste tops often find their way into the drains of lavatories. Hair is another major cause of blockages in lavatories. Avoid these problems by being aware of what goes down the drain. Keep the pop-up plug in the lavatory drain. The plug will prevent large objects from entering the trap and pipes.

Laundry Tubs

Keep a screen over the drain of your laundry tub. Lint and dirt are the largest causes of trouble with laundry tubs. These objects enter the drain and block the drainage pipes.

Washer Hookups

Check your washing machine hoses every few months. If the hoses appear cracked or brittle, replace them. If a washing machine hose breaks while you are away, the entire area can flood. Washing-machine hoses are responsible for the flooding of many homes each year.

Water Heater Relief Valves

Test the relief valve on your water heater every now and then. Lift the lever and let the relief valve blow off a little water. This safety valve is very important. By testing the relief valve, you reduce the risk of having your water heater explode.

Shut-Off Valves

Check your shut-off valves at least twice a year to be sure they work. In the midst of a plumbing emergency is no time to discover your shut-off valves don't work.

Kitchen Sinks

Use common sense in what you let drain through your kitchen sink. Grease is the biggest problem in kitchen sink drains. Never pour oil or grease into the kitchen sink. Avoid letting large food particles wash down the drain.

COMMON SENSE

Common sense is your best friend when taking preventive measures with your plumbing. Think about what you are doing and what effect it will have on your plumbing. Read and follow the instructions supplied with your plumbing products. Plumbing does not require you to babysit it, but it does work better when treated with respect.

By now you know enough about your home's plumbing to decide what your limitations are. Never gamble with your plumbing. If you don't understand how to work on a portion of your plumbing, call a professional plumber. Plumbers love it when homeowners work on their own plumbing. The plumber knows that in many cases, the homeowner will have to call for help. By the time the plumber arrives, the homeowner has often caused enough damage for the plumber to make more money than if he had been called before the homeowner's attempt to work with the plumbing.

With the help of this book, you don't have to be one of those hapless homeowners. You should be able to recognize when you might be getting in over your head. As I said early in the book, safety should be your primary concern. Never attempt any of these repairs or replacements without the proper tools and safety gear. Good luck with all your plumbing attempts.

Glossary

ABS — Acrylonitrile Butadiene Styrene. This is normally used to describe schedule 40 plastic pipe. The pipe is black.

ACCESSIBLE — To have access to a part of the plumbing system. If you are required to provide access, you may install an access panel or some other device that must be removed before access is gained to the plumbing.

ADAPTER — An approved fitting that allows the mating of different types of materials or fittings.

AIR CHAMBER — A device installed in the potable water system to reduce the effect of air hammer. It gives additional room in the piping system for air to compress and water to expand.

AIR GAP (distance) — The unobstructed vertical distance any device conveying water or waste has between its location and the flood level rim at the fixture or other device receiving the water or waste.

AIR GAP (product) — The device used to handle drainage from a dishwasher. It is the item that normally sits near the kitchen faucet and is connected to the drain hose from the dishwasher to the sanitary drainage system.

ANGLE STOP — The cut-off valve frequently found where the water supply pipes come out of the wall.

ANTI-SIPHON — A valve or other device that eliminates the risk of siphonage.

ANTI-SIPHON VALVE — A valve designed to eliminate siphonage back through the valve.

AUTOMATIC VENT — A mechanical vent used to vent a portion of the drainage system. These vents are usually made of plastic and contain a diaphragm. The vents are screwed into a female adapter and provide air to the drainage system. As a fixture's drain fills with water, the diaphragm is pulled down to open the vent, allowing air into the drain pipe. This air causes the fixture to drain faster. When used, they are normally used in remodeling jobs; they are not approved for use without special permission from the code enforcement office.

BACK VENT — The vent extending from the drainage of a fixture. Back vents are commonly dry vents, extending above the fixture's flood level rim. They sometimes tie into other vents or may run independently to outside air.

BACKFLOW — The reverse flow of water in the plumbing system.

BACKFLOW PREVENTER — A device used to prevent the danger of backflow.

BACK-SIPHONAGE—A condition in which water flows backwards in a plumbing system. It is caused by negative pressure in the pipe.

BACKWATER VALVE — A valve used in the drainage system. It prevents sewage from flowing backwards into the building and escaping through the plumbing fixtures.

BALLCOCK — The device used in filling a toilet with water. It is operated with a float, to supply water to a fixture.

BELL-AND-SPIGOT CAST IRON PIPE — A heavy cast iron pipe used in the dwv system. This type of pipe has a bell or hub on one end. The other end of the pipe is straight. To make a connection between two pieces of pipe, the straight end of one pipe is inserted in the hub of the other pipe. The watertight connection is made with oakum and molten lead or rubber devices that fit inside the hub, prior to the insertion of the pipe.

BIDET — A personal hygiene fixture. While not common in most homes, bidets are found in many expensive homes.

BLEED — To purge air from a system. For example, you might bleed the air from the pipes supplying your water pump with water.

BRANCH — A pipe that is a part of the plumbing system. Branches originate from the water main or building drain and extend out to a fixture at some distance from the primary pipe. A branch could be any part of the plumbing system that is not a riser, main, or stack.

BRANCH VENT — Connects an individual vent with a main vent stack or stack vent. A branch vent can serve a single individual vent or multiple individual vents.

BUILDING DRAIN — The primary pipe carrying drainage through a building. When the building drain exits the building, it becomes the building sewer.

BUILDING SEWER — Conveys waste from the termination of the building drain to the connection of the municipal sewer or private waste disposal system.

BUSHING — A fitting that fits inside the hub of another fitting to reduce its size or alter its interior. For example, a 3" x 2" bushing would be installed in the hub of a 3-inch fitting to allow the fitting to accept a 2-inch pipe.

CAP — A fitting placed over the end of the pipe to close the pipe.

CAULKING (cast iron) — The process of making joints watertight with molten lead.

CAULKING (fixtures) — The act of sealing around fixtures with a sealant to prevent water penetration.

CESSPOOL — A hole in the ground used to accept the discharge of a drainage system. The hole is lined to retain solids and organic material, while allowing liquids to pass through, into the earth.

CHECK VALVE — A device used to ensure that the contents of a pipe are only allowed to flow in one direction. A common use of check valves includes installation on the pipes of pumps.

CIRCUIT VENT — A vent that serves multiple fixtures. It extends from the low end of the highest fixture connection of a horizontal branch to the vent stack.

CLEANOUT — An accessible opening in the dwv system that allows the cleaning of the pipes with sewer machines and snakes.

CLOSET AUGER — A device used to remove blockages in the traps of toilets. It is a hand-operated tool with a spring head and curved rod, made to negotiate the tight turns of a toilet drain.

CLOSET BEND — A quarter-bend or 90° elbow. Closet bends may have a 3-inch opening on one end of the ell and a 4-inch opening on the other end of the ell.

COMMON VENT — A vent that serves multiple fixtures.

COMPRESSION FITTING — A device using ferrules and nuts on the body of a coupling or fitting to make watertight connections. Compression fittings are frequently found where the supply tubes of faucets enter the cut-off valve.

CONTINUOUS VENT — A vent that is a continuation of the drain it serves. Continuous vents are vertical vents and may be referred to as back vents.

CONTINUOUS WASTE — The piping used to connect multiple drains to a common trap. Most double-bowl kitchen sinks are equipped with a continuous waste.

COUPLING — A fitting allowing the connection of two pipes to form a continuous run of piping.

CPVC — Chlorinated Polyvinyl Chloride plastic pipe. This is the rigid plastic pipe used in potable water systems.

CRITICAL LEVEL — The term used to describe the level at which a vacuum breaker may be submerged before backflow occurs.

CROSS CONNECTION — A cross connection exists when two separate piping systems, such as the hot and cold water pipes, are allowed to mingle their contents with each other. Cross connections could occur at a faucet, washing machine, or some other location.

CROWN-VENTED TRAP — A trap that has its vent extending upwards from the top of the trap, rather than from the trap arm.

DEVELOPED LENGTH — The measurement or distance of all piping installed. To calculate the developed length of piping, you must measure each piece of pipe.

DRAIN — Any pipe that carries waste in a plumbing system.

DRAINAGE FITTING — Any fitting used in the drainage portion of a plumbing system.

DRUM TRAP — A trap that will not allow the back-siphonage of the trap, even without a vent. Drum traps are not self-cleaning and are generally prohibited, without special permission from the code enforcement office.

DRYWELL — A drywell is sometimes used to receive the discharge from surface-water drains. A drywell is typically lined with stone to accept the water and to allow it to soak into the soil.

DWV — Abbreviation for the drain, waste, and vent system.

ELBOW — A plumbing fitting for either water distribution or the dwv system. It may be called an elbow, an ell, a quarter-bend, or a ninety.

ESCUTCHEON — The chrome ring found around a pipe penetrating a finished wall or floor. The escutcheon provides a neat and finished look to the installation of pipes, and it prevents rodents from climbing the pipes to enter the premises.

FAUCET — The device that controls the flow and mix of water entering a sink, lavatory, tub, shower, or other plumbing fixture.

FEMALE CONNECTION — A connection with interior threads, designed to accept the external threads of a male adapter.

FILTER — A device used to filter substances. The aerator on your faucet is a form of filter.

FINISH PLUMBING — The setting of fixtures, such as toilets, sinks, and faucets.

FITTING — Used to identify the parts of the plumbing system used to connect the piping.

FIXTURE BRANCH — The water supply pipes running between the primary water pipes and the supply tubes of individual plumbing fixtures.

FIXTURE DRAIN — The section of piping that runs from the fixture's trap to the connection with the dwv system. Fixture drains are often called trap arms.

FIXTURE SUPPLY — The pipe that runs between a fixture and the fixture branch.

FIXTURE UNIT — A unit of measure used to calculate the load and demands of fixtures on a plumbing system.

FLANGE — A mounting surface. The most frequent use of the word in plumbing refers to closet flanges, the devices mounted to the floor to receive and hold toilet bowls.

FLAPPER — A rubber device used to seal the opening of a flush valve in a toilet tank.

FLARE FITTING — A fitting made to be used with flared piping. Flare fittings are specially formed to mate with flared pipe in creating a leakproof joint.

FLASHING — The device used to seal around vent terminals on the roof of a structure.

FLEX CONNECTOR — Usually a short piece of material used to connect a device with the piping of a plumbing or gas piping system. Flex connectors are flexible and allow the device connected to the rigid piping to move, without putting stress on the rigid piping.

FLOAT BALL — The float on the end of the float rod in the toilet tank. It is the float that operates the ballcock.

FLOAT ROD — A float rod spans the distance between a ballcock and a float ball. The float rod screws into the ballcock and the float ball to allow proper operation of the ballcock.

FLOOD-LEVEL RIM — The point where water will flood out of a fixture. For example, the edge of your kitchen sink, where it meets the countertop, is the flood-level rim of the sink.

FLUSH VALVE — The device in a toilet tank that allows water to pass from the toilet tank to the toilet bowl.

FLUX — A substance applied to copper pipe when the pipe will be soldered. The flux acts as a cleaning agent to ensure a good solder joint.

GATE VALVE — A valve that uses a forged metal gate to close the valve. Instead of using a rubber washer, which may deteriorate, gate valves close with the use of the gate, ensuring a more positive closing of the valve.

GRADE — Term used to describe the slope, fall, or pitch of a pipe.

HOSE BIBB — A device used to supply water to a garden hose. Hose bibbs accept a water supply pipe on one end and the threads of a garden hose on the other end. They are equipped with a handle and an assembly to control the flow of water.

HOT WATER — In terms of plumbing, water is considered to be hot water when its temperature is at or above 110° F.

HOUSE TRAP — House traps are not allowed by present plumbing codes. House traps can still be found in existing buildings. A house trap is a trap installed in the building drain, just before the building drain exits the building.

HUB — The part of a pipe or fitting designed to accept the end of a piece of pipe.

HUBLESS CAST IRON — A lightweight cast iron without hubs. The connections made with this type of pipe are done with special bands and clamps.

HYDROSTATIC TEST — A test of the plumbing system using water. If you test your waste or water lines with water, you are performing a hydrostatic test.

INDIRECT WASTE PIPE — A pipe that does not connect directly to the sanitary drainage system. Instead, it conveys its contents to the sanitary drainage system through an air gap, to a fixture or other receptacle.

INDIVIDUAL VENT — A vent serving only one trap.

J-BEND — The portion of a trap that retains water at all times.

JOINT — Refers to connections made in the installation of plumbing.

LONG-SWEEP FITTING — A fitting with a more gradual turn than its short-turn counterpart. The more gradual bend reduces the risks of pipe blockages.

MAIN — Used to describe a primary pipe, such as a water main.

MALE ADAPTER — A fitting with exposed threads, designed to screw into a female adapter.

NIPPLE — A short piece of pipe with threads on each end.

NON-POTABLE WATER — Water that is not safe to drink.

OAKUM — A material packed in the hub of cast iron pipe, surrounding the pipe extending into the hub, before molten lead is poured into the hub to make a joint.

OFFSET — A change in direction. Anytime piping is turned in a different direction with the use of fittings, an offset is created.

O-RING — A rubber ring used in plumbing parts to seal them against water leakage.

PACKING NUT — The packing nut of a valve is

the nut that holds the packing in the valve to prevent leaks.

PB PIPE — Polybutylene pipe. It is a very flexible plastic pipe used for water distribution.

PE PIPE — Polyethylene pipe. It is a plastic pipe that comes in rolls and is frequently used for water service applications outside the building.

PIPE JOINT COMPOUND — Also called pipe dope, pipe joint compound is a sealant applied to threads before making a screw connection.

PITCH — See GRADE.

PLUG — Used for the same purpose as a cap, but plugs are screwed into fittings instead of being installed over pipes.

PLUMBING — The trade or work pertaining to the installation, repair, alteration, and removal of plumbing and drainage systems.

PLUMBING CODE — A set of rules, regulations, and/or laws that dictate the manner in which the trade of plumbing may be conducted.

PLUMBING OFFICIAL — Also plumbing inspector. This is the authorized person designated to inspect and enforce the plumbing code.

PLUMBING SYSTEM — A plumbing system includes all water distribution pipes, waste and vent pipes, plumbing fixtures, traps, and devices used for plumbing within the property lines of the premises.

POTABLE WATER — Water that is safe for drinking, cooking, and domestic use.

PRESSURE PIPE — A pipe meant to handle contents that are under pressure, such as a water distribution pipe.

PRESSURE-REDUCING VALVE — Used to govern the pressure of a substance entering a pipe. The most common use of pressure-reducing valves is on the water main of a building. If the water pressure produced by a municipal source is too high for the building's use, a pressure-reducing valve is installed to lower the pressure entering the building's water distribution system.

P-TRAP — The most common type of trap used in modern plumbing. It is used for trap arms that come out of the wall, as opposed to the floor.

PVC — Polyvinyl Chloride plastic pipe. This pipe is frequently used in dwv systems. When used for this purpose, the PVC pipe is white schedule 40 plastic pipe.

RECEPTOR — An approved device intended to accept the discharge from an indirect waste.

REDUCER — Plumbers refer to fittings that reduce the size of a pipe or fitting as reducers. Unlike bushings, reducers are installed over pipe, rather than in the hub of fittings.

RELIEF-VALVE DRAIN — The drain that handles the discharge of a relief valve. These drains usually terminate into open air, not into the sanitary drainage system.

RISER — A pipe that rises vertically.

ROUGH-IN — The plumbing installed prior to the completion of walls, ceilings, and floors. If you were plumbing a new house, the plumbing you installed before drywall was placed on the walls and ceilings would be rough-in plumbing.

SADDLE VALVE — A valve used to tap into existing pipe. A common use of a saddle valve would be to supply an icemaker with water from the supply pipe of a kitchen faucet.

SANITARY FITTING — A fitting used in drainage piping.

SANITARY PLUMBING — The plumbing that removes waste from a building.

SCHEDULE 40 PLASTIC — Plastic pipe usually used for drains and vents. Schedule 40 is a rating that refers to the wall thickness of the pipe.

SEPTIC TANK — Part of a private waste disposal system. A septic tank accepts the discharge from a building sewer and holds it for distribution into the septic field.

SEWAGE — Any liquid waste containing animal or vegetable matter.

SLIP FITTING — A fitting that will slide over a pipe and continue to slide down the pipe. Slip couplings are the type of slip fitting most often used. They allow the coupling to slide back onto the pipe so that the couplings can be used with a minimum of space. By using slip couplings, you do not need to gain much movement of the pipe to make a joint.

SOIL PIPE — A pipe that transports sewage containing fecal matter.

SOIL STACK — A soil pipe that extends vertically to accept the discharge of toilets.

SOLDER — Used to make watertight joints with copper pipe and fittings.

SOLDERED JOINT — A joint made with solder.

SOLVENT CEMENT — The glue used to make connections with various plastic pipes.

STANDPIPE — Standpipes are most often encountered as receptors for the drainage conveyed by washing machines. They are the vertical receptor extending from a trap to accept indirect waste.

S-TRAP — S-traps are now illegal. These traps are used on fixtures when the drain for the fixture comes up through the floor, instead of coming out of a wall.

STREET FITTING — A fitting made so that one end of the fitting will fit into the hub of another fitting.

SUMP — A tank or basin that receives sewage or water to be pumped to another location.

SUMP PUMP — Used to pump water from a sump. It is not intended to pump sewage.

SUMP VENT — When a sump is used to receive sewage, the sump should have a sump vent. The sump vent controls odor and sewer gas that builds up in the sump.

SWEATING (condensation) — It is common for people to say their toilet tank is sweating. This means condensation is building on the outside of the tank.

SWEATING (soldering) — Plumbers often refer to the act of soldering as sweating. They will say they are sweating the copper pipes and fittings, to mean they are soldering the pipe and fittings.

TAILPIECE — The tubing that extends from a fixture's drain to the fixture's trap.

TANK BALL — Used to seal the openings of some flush valves, in toilet tanks.

TEE — A fitting that allows a secondary pipe to branch off of a main pipe.

TEST TEE — A special fitting used in the dwv system. A tee has a flat face with the tee portion being threaded to accept a clean-out plug. A test tee allows the testing of the dwv system when it has already been connected to the main sewer.

T & P VALVE — A temperature and pressure-relief valve. These valves can be found on water heaters. They are a safety device to protect against excessive heat or pressure in the water heater.

TRAP — A device used to prevent sewer gas from entering the atmosphere of a building.

TRAP ARM — The section of drainage pipe that runs from the trap to the connection with the building's dwv system.

TRAP SEAL — The seal made by water that remains in the trap at all times. Trap seals prevent sewer gas from entering the atmosphere inside the building.

TUBING — Small, usually flexible, piping. Tubing can be rigid. Technically, most copper pipe used to plumb a home is tubing. This is rigid tubing, but most people refer to it as pipe.

UNION — A fitting used to couple two pipes together. Unions are machined to fit together without leaking. When unions are installed, the connection can be loosened and the two pipes can be separated without cutting the pipe.

VACUUM — A pressure less than that exerted by the atmosphere.

VACUUM BREAKER — A fitting that protects against backflow on an opening under normal atmospheric pressure.

VENT PIPE — A pipe used to vent a plumbing fixture.

VENT STACK — A vertical vent pipe connected to the drainage system.

VENT SYSTEM — Designed to provide air circulation to a drainage system and to prevent the siphonage of traps.

WASTE — The discharge from a plumbing device that does not contain fecal matter.

WASTE PIPE — A pipe conveying waste without fecal matter.

WATER CLOSET — A toilet.

WATER DISTRIBUTION PIPE — Pipe that distributes potable water to plumbing fixtures.

WATER HAMMER — A pressure surge within the water distribution system.

WATER HEATER — A device used to heat cold water to a temperature of at least 110° F.

WATER MAIN — A water supply pipe for public use.

WATER METER — A device installed on a water service to measure the amount of water used by a plumbing system.

WATER SOFTENER — A device used to condition water and to remove hardness from water.

WATER SUPPLY PIPE — Also called a water service. It is the pipe supplying water from the main source to a building.

WATER SUPPLY SYSTEM — The water supply system is composed of all the pipes and parts necessary to supply water to a building.

WAX RING — Used to create a seal between a toilet bowl and its flange.

WELL — A source of water.

WELL PUMP — A pump designed to pump water from a well to the water distribution system of a building.

WET VENT — A vent that receives the discharge of a drain.

WING ELBOW — Also known as a drop-eared ell. It is an elbow with two ears on it. The ear extensions allow the ell to be secured with screws or nails to wood blocking. The most common use of these fittings is to accept the end of a shower head's arm.

WYE — A fitting that allows a branch pipe to enter a primary pipe on a gradual angle.

Appendices

1. Troubleshooting Checklists

TOILET			
Symptom	**Probable Cause**	**Symptom**	**Probable Cause**
Will not flush	No water in tank Stoppage in drainage system	Water runs constantly	Bad flapper or tank ball Bad ballcock Float rod needs adjusting Float is filled with water Ballcock needs adjusting Pitted flush valve Undiscovered leak Cracked overflow tube
Flushes poorly	Clogged flush holes Flapper or tank ball is not staying open long enough Not enough water in tank		
	Partial drain blockage Defective handle Bad connection between handle and flush valve Vent is clogged	Water seeps from base of toilet	Bad wax ring Cracked toilet bowl
		Water dripping from tank	Condensation Bad tank-to-bowl gasket Bad tank-to-bowl bolts Cracked tank Flush-valve nut is loose
Water droplets covering tank	Condensation		
Tank fills slowly	Defective ballcock Obstructed supply pipe Low water pressure Partially closed valve Partially frozen pipe	No water comes into the tank	Closed valve Defective ballcock Frozen pipe Broken pipe
Makes unusual noises when flushed	Defective ballcock		

LAVATORY			
Symptom	**Probable Cause**	**Symptom**	**Probable Cause**
Faucet drips from spout	Bad washers or cartridge Bad faucet seats	Drains slowly	Hair on pop-up assembly Partial obstruction in drain or trap Pop-up needs to be adjusted
Faucet leaks at base of spout	Bad "O" ring		
Faucet will not shut off	Bad washers or cartridge Bad faucet seats	Will not drain	Blocked drain or trap Pop-up is defective
Poor water pressure	Partially closed valve Clogged aerator Not enough water pressure Blockage in the faucet Partially frozen pipe	Gurgles as it drains	Partial drain blockage Partial blockage in the vent
		Won't hold water	Pop-up needs adjusting Bad putty seal on drain
No water	Closed valve Broken pipe Frozen pipe		

BATHTUB			
Symptom	**Probable Cause**	**Symptom**	**Probable Cause**
Won't drain	Clogged drain Clogged tub waste Clogged trap	Water comes out spout and shower at the same time	Bad diverter washer Bad diverter seat Bad diverter
Drains slowly	Hair in tub waste Partial drain blockage	Faucet will not shut off	Bad washers or cartridge Bad faucet seats
Won't hold water	Tub waste needs adjusting	Poor water pressure	Partially closed valve Not enough water pressure Blockage in the faucet Partially frozen pipe
Won't release water	Tub waste needs adjusting		
Gurgles as it drains	Partial drain blockage Partial blockage in the vent	No water	Closed valve Broken pipe Frozen pipe
Water drips from spout	Bad faucet washers/ cartridge Bad faucet seats		

SHOWER

Symptom	Probable Cause	Symptom	Probable Cause
Won't drain	Clogged drain Clogged strainer Clogged trap	Faucet will not shut off	Bad washers or cartridge Bad faucet seats
Drains slowly	Hair in strainer Partial drain blockage	Poor water pressure	Partially closed valve Not enough water pressure
Gurgles as it drains	Partial drain blockage Partial blockage in the vent		Blockage in the faucet Partially frozen pipe
Water drips from shower head	Bad faucet washers/ shower head cartridge Bad faucet seat	No water	Closed valve Broken pipe Frozen pipe

KITCHEN SINK

Symptom	Probable Cause	Symptom	Probable Cause
Faucet drips from spout	Bad washers or cartridge Bad faucet seats	Drains slowly	Partial obstruction in drain or trap
Faucet leaks at base of spout	Bad "O" ring	Will not drain	Blocked drain or trap
		Gurgles as it drains	Partial drain blockage Partial blockage in the vent
Faucet will not shut off	Bad washers or cartridge Bad faucet seats		
Poor water pressure	Partially closed valve Clogged aerator Blockage in the faucet Partially frozen pipe	Won't hold water	Bad basket strainer Bad putty seal on drain
		Spray attachment will not spray	Clogged holes in spray head Kinked spray hose
No water	Closed valve Broken pipe Frozen pipe	Spray attachment will not cut off	Bad spray head

LAUNDRY TUB			
Symptom	**Probable Cause**	**Symptom**	**Probable Cause**
Faucet drips from spout	Bad washers or cartridge Bad faucet seats	No water	Closed valve Broken pipe Frozen pipe
Faucet leaks at base of spout	Bad "O" ring	Drains slowly	Partial obstruction in drain or trap
Faucet will not shut off	Bad washers or cartridge Bad faucet seats	Will not drain	Blocked drain or trap
Poor water pressure	Partially closed valve Clogged aerator Not enough water pressure Blockage in the faucet Partially frozen pipe	Gurgles as it drains	Partial drain blockage Partial blockage in the vent
		Won't hold water	Bad basket strainer Bad putty seal on drain

ELECTRIC WATER HEATER			
Symptom	**Probable Cause**	**Symptom**	**Probable Cause**
Relief valve leaks slowly	Bad relief valve	Too little hot water	An element is bad Bad thermostat Thermostat needs adjusting
Relief valve blows off periodically	High water temperature High pressure in tank Bad relief valve	Too much hot water	Thermostat needs adjusting Controls are defective
No hot water	Electrical power is off Elements are bad Defective thermostat Inlet valve is closed	Water leaks from tank	Hole in tank Rusted-out fitting in tank

GAS WATER HEATER

Symptom	Probable Cause	Symptom	Probable Cause
Relief valve leaks slowly	Bad relief valve	Too little hot water	Bad thermostat Thermostat needs adjusting
Relief valve blows off periodically	High water temperature High pressure in tank Bad relief valve	Too much hot water	Thermostat needs adjusting Controls are defective Burner will not shut off
No hot water	Out of gas Pilot light is out Bad thermostat Control valve is off Gas valve closed	Water leaks from tank	Hole in tank Rusted-out fitting in tank

SUBMERSIBLE POTABLE-WATER PUMP

Symptom	Probable Cause	Symptom	Probable Cause
Won't start	No electrical power Wrong voltage Bad pressure switch Bad electrical connection	Runs, but does not produce water, or only produces a small quantity	Check valve stuck in closed position Check valve installed backwards Bad electrical wiring Wrong voltage Pump is sitting above the water in the well Leak in the piping Bad pump or motor Broken pump shaft Clogged strainer Jammed impeller
Starts, but shuts off fast	Circuit breaker or fuse is inadequate Wrong voltage Bad control box Bad electrical connections Bad pressure switch Pipe blockage Pump is seized Control box is too hot	Pump runs too often	Check valve stuck open Pressure tank is waterlogged and needs air injected Pressure switch needs adjusting Leak in piping Wrong size pressure tank
Low water pressure in pressure tank	Pressure switch needs adjusting Bad pump Leak in piping Wrong voltage		

JET POTABLE-WATER PUMP			
Symptom	**Probable Cause**	**Symptom**	**Probable Cause**
Won't start	No electrical power Wrong voltage Bad pressure switch Bad electrical connection Bad motor Motor contacts are open Motor shaft is seized	Starts and stops too often	Leak in the piping Bad pressure switch Bad air control valve Waterlogged pressure tank Leak in pressure tank
Runs, but produces no water	Needs to be primed Foot valve is above the water level in the well Strainer clogged Suction leak	Low water pressure in pressure tank	Strainer on foot valve is partially blocked Leak in piping Bad air charger Worn impeller hub Lift demand is too much for the pump
Pump does not cut off when working pressure is obtained	Pressure switch is bad Pressure switch needs adjusting Blockage in the piping		

2. Material Take-Off Checklists

TOILET REPAIRS			
ITEM	HAVE	NEED	DON'T NEED
Wax ring	____	____	____
Closet bolts	____	____	____
Stop valve	____	____	____
Compression nuts	____	____	____
Compression ferrules	____	____	____
Supply tube	____	____	____
Ballcock	____	____	____
Flush valve	____	____	____
Float rod	____	____	____
Float	____	____	____
Flapper/tank ball	____	____	____
Tank-to-bowl bolts	____	____	____
Tank-to-bowl gasket	____	____	____
Pipe joint compound	____	____	____
Plumber's putty	____	____	____

BATHTUB AND SHOWER REPAIRS			
ITEM	HAVE	NEED	DON'T NEED
Trap	_____	_____	_____
Tub waste and overflow	_____	_____	_____
Slip-nuts/washers	_____	_____	_____
Tub spout	_____	_____	_____
Shower head	_____	_____	_____
Shower arm	_____	_____	_____
New faucet	_____	_____	_____
Faucet stems	_____	_____	_____
Faucet washers	_____	_____	_____
Faucet cartridge	_____	_____	_____
Faucet seats	_____	_____	_____
"O" rings	_____	_____	_____
Packing material	_____	_____	_____
Pipe joint compound	_____	_____	_____
Plumber's putty	_____	_____	_____
Lead-free solder	_____	_____	_____
Flux	_____	_____	_____
Pipe	_____	_____	_____
Pipe fittings	_____	_____	_____

LAVATORY REPAIRS

ITEM	HAVE	NEED	DON'T NEED
Trap	____	____	____
Slip-nuts/washers	____	____	____
Pop-up assembly	____	____	____
Stop valves	____	____	____
Supply tubes	____	____	____
Compression nuts	____	____	____
Compression ferrules	____	____	____
New faucet	____	____	____
Faucet stems	____	____	____
Faucet washers	____	____	____
Faucet cartridge	____	____	____
Faucet seats	____	____	____
Faucet aerator	____	____	____
"O" rings	____	____	____
Packing material	____	____	____
Pipe joint compound	____	____	____
Plumber's putty	____	____	____
Lead-free solder	____	____	____
Flux	____	____	____
Pipe	____	____	____
Pipe fittings	____	____	____

KITCHEN SINK REPAIRS			
ITEM	HAVE	NEED	DON'T NEED
Trap	———	———	———
Continuous waste	———	———	———
Slip-nuts/washers	———	———	———
Basket strainers	———	———	———
Air gap	———	———	———
Hose clamps	———	———	———
Rubber hose	———	———	———
Stop valves	———	———	———
Supply tubes	———	———	———
Compression nuts	———	———	———
Compression ferrules	———	———	———
Spray head/hose	———	———	———
New faucet	———	———	———
Faucet stems	———	———	———
Faucet washers	———	———	———
Faucet cartridge	———	———	———
Faucet seats	———	———	———
Faucet aerator	———	———	———
"O" rings	———	———	———
Packing material	———	———	———
Pipe joint compound	———	———	———
Plumber's putty	———	———	———
Lead-free solder	———	———	———
Flux	———	———	———
Pipe and fittings	———	———	———

LAUNDRY SINK REPAIRS

ITEM	HAVE	NEED	DON'T NEED
Trap	____	____	____
Slip-nuts/washers	____	____	____
Basket strainer	____	____	____
Stop valves	____	____	____
Supply tubes	____	____	____
Compression nuts	____	____	____
Compression ferrules	____	____	____
New faucet	____	____	____
Faucet stems	____	____	____
Faucet washers	____	____	____
Faucet cartridge	____	____	____
Faucet seats	____	____	____
Faucet aerator	____	____	____
"O" rings	____	____	____
Packing material	____	____	____
Pipe joint compound	____	____	____
Plumber's putty	____	____	____
Lead-free solder	____	____	____
Flux	____	____	____
Pipe and fittings	____	____	____

WASHING MACHINE HOOKUP REPAIRS			
ITEM	HAVE	NEED	DON'T NEED
Trap	_____	_____	_____
Stop valves	_____	_____	_____
Washing machine hoses	_____	_____	_____
Drain hose	_____	_____	_____
Hose clamps	_____	_____	_____
Compression nuts	_____	_____	_____
Compression ferrules	_____	_____	_____
New faucet	_____	_____	_____
Faucet stems	_____	_____	_____
Faucet washers	_____	_____	_____
Faucet cartridge	_____	_____	_____
Faucet seats	_____	_____	_____
Faucet aerator	_____	_____	_____
"O" rings	_____	_____	_____
Packing material	_____	_____	_____
Pipe joint compound	_____	_____	_____
Plumber's putty	_____	_____	_____
Lead-free solder	_____	_____	_____
Flux	_____	_____	_____
Pipe and fittings	_____	_____	_____

DRAINAGE FITTINGS NEEDED			
FITTING	**TYPE**	**SIZE**	**QUANTITY**
Wye	_____	_____	_____
Combination	_____	_____	_____
Sanitary tee	_____	_____	_____
Long sweep quarter bend	_____	_____	_____
Short sweep quarter bend	_____	_____	_____
Fifth-bend	_____	_____	_____
Sixth-bend	_____	_____	_____
Sixteenth-bend	_____	_____	_____
Wye and eighth bend	_____	_____	_____
Sanitary tapped tee	_____	_____	_____
Closet flange	_____	_____	_____
Trap	_____	_____	_____
Plug	_____	_____	_____
Coupling	_____	_____	_____
Slip couplings	_____	_____	_____
Bushing	_____	_____	_____
Reducer	_____	_____	_____
Reducing tee	_____	_____	_____
Reducing wye	_____	_____	_____
Male adapter	_____	_____	_____
Female adapter	_____	_____	_____
Clean-out	_____	_____	_____
Clean-out plug	_____	_____	_____
Street fittings	_____	_____	_____
Other	_____	_____	_____

WATER PIPE FITTINGS NEEDED			
FITTING	TYPE	SIZE	QUANTITY
Tee	_____	_____	_____
Coupling	_____	_____	_____
Union	_____	_____	_____
Reducer	_____	_____	_____
Male adapter	_____	_____	_____
Female adapter	_____	_____	_____
90° elbow	_____	_____	_____
45° elbow	_____	_____	_____
Drop-eared elbow	_____	_____	_____
Reducing tee	_____	_____	_____
Reducing elbow	_____	_____	_____
Insert fittings	_____	_____	_____
Street fittings	_____	_____	_____
Caps	_____	_____	_____

3. Plumbing Charts and Tables

RECOMMENDED TRAP SIZES

TYPE OF FIXTURE	TRAP SIZE IN INCHES
Washing machine	2
Laundry tub	$1^1/_2$
Bathtub	$1^1/_2$
Shower	2
Lavatory	$1^1/_4$
Bidet	$1^1/_4$
Kitchen sink	$1^1/_2$
Bar sink	$1^1/_2$
Floor drain	2

BUILDING DRAINS AND SEWERS

NUMBER OF FIXTURES ALLOWED ON A SEWER OR BUILDING DRAIN

(The fixture unit amounts are based on a pipe with a quarter of an inch to the foot in pitch.)

PIPE SIZE IN INCHES	MAXIMUM FIXTURE UNITS ALLOWED
2	21
3	42 (not more than two toilets)
4	216

DRAINAGE FIXTURE UNITS

TYPE OF FIXTURE	FIXTURE UNIT VALUE
Washing machine	3
Laundry tub	2
Bathtub	2
Shower	2
Lavatory	1
Bidet	1
Toilet	4
Kitchen sink	2
Bar sink	1
Floor drain	2

MINIMUM PITCH REQUIRED ON HORIZONTAL DRAINS

PIPE SIZE IN INCHES	MINIMUM PITCH (inch per foot)
$1^1/_2$	$1/_4$
2	$1/_4$
3	$1/_8$
4	$1/_8$

MINIMUM SIZING FOR SEWER PUMPS

DISCHARGE PIPE SIZE IN INCHES	GALLONS-PER-MINUTE OF PUMP
2	21
$2^1/_2$	30
3	46

AVERAGE WATER USAGE OF FIXTURES IN GALLONS PER MINUTE

FIXTURE	FLOW RATE (gpm)
Bathtub	4
Washing machine	4
Dishwasher	3
Kitchen sink	$2^1/_2$
Laundry tub	4
Lavatory	2
Shower	3
Toilet	3

SIZING FOR MINIMAL WATER SUPPLY REQUIREMENTS

FIXTURE	MINIMUM PIPE SIZE IN INCHES
Bathtub	$1/_2$
Bidet	$3/_8$
Dishwasher	$1/_2$
Hose bibb	$1/_2$
Kitchen sink	$1/_2$
Laundry tub	$1/_2$
Lavatory	$3/_8$
Shower	$1/_2$
Toilet (two-piece)	$3/_8$
Toilet (one-piece)	$1/_2$

MAXIMUM DISTANCE FROM A FIXTURE TRAP TO ITS VENT			
TRAP SIZE (inches)	DRAIN SIZE (inches)	PITCH (inch per foot)	DISTANCE ALLOWED (feet)
1¼	1¼	¼	3½
1¼	1½	¼	5
1½	1½	¼	5
1½	2	¼	8
2	2	¼	6
3	3	⅛	10
4	4	⅛	12

WATER PIPE FIXTURE UNIT VALUES			
FIXTURE	HOT	COLD	TOTAL UNIT VALUES
Bathtub	1½	1½	2 (when combined)
Bidet	1½	1½	2 (when combined)
Kitchen sink	1½	1½	2 (when combined)
Laundry tub	2	2	3 (when combined)
Lavatory	1½	1½	2 (when combined)
Shower head	3	3	4 (when combined)
Toilet	0	5	5

4. Resource Guide for Plumbing Products and Tools

American Standard Inc.
Plumbing Products Group
One Centennial Plaza
P.O. Box 6820
Piscataway, NJ 08855-6820

A.O. Smith Water Products Co.
P.O. Box 1499
Camden, SC 29020

Aurora Pump
800 Airport Rd.
North Aurora, IL 60542-9977

Autotrol
5730 North Glen Park Rd.
Milwaukee, WI 53209

A.W. Cash Valve Mfg. Corp.
666 E. Wabash
Decatur, IL 62525

Central Brass
2950 East 55th St.
Cleveland, OH 44127

Chemical Engineering Corp.
P.O. Box 266
Churubusco, IN 46723

The Chicago Faucet Co.
2100 South Nuclear Dr.
Des Plaines, IL 60018

Conbraco
P.O. Box 247
Matthews, NC 28106

Eljer
3 Gateway Center
Pittsburgh, PA 15222

Fernco
300 S. Dayton St.
Davison, MI 48423

Fiat Products
One Michael Ct.
Plainview, NY 11803

General Wire Spring Co.
1101 Thompson Ave.
McKees Rocks, PA 15136

Gerber Plumbing Fixtures
4656 West Touhy Ave.
Chicago, IL 60646

Goulds Pumps, Inc.
P.O. Box 68
East Bayard St.
Seneca Falls, NY 13148

Guy Gray Manufacturing Co.
P.O. Box 2287
Paducah, KY 42002-2287

Kohler Co.
Kohler, WI 53044

Leonard Valve Co.
1360 Elmwood Ave.
Cranston, RI 02910

Little Giant Pump Co.
3810 N. Tulsa
Oklahoma City, OK 73112

Microphor, Inc.
Plumbing Division
P.O. Box 1460
452 East Hill Rd.
Willits, CA 95490-1460

Midwest Supply Co.
One South Park Rd.
Joliet, IL 60433

E.L. Mustee & Sons, Inc.
5431 West 164th St.
Cleveland, OH 44142

F.E. Myers
1101 Myers Parkway
Ashland, OH 44805-1923

Price Pfister
13500 Paxton St.
Pacoima, CA 91331

Rain Soft
2080 E. Lunt Ave.
Elk Grove Village, IL 60007

Republic Products
P.O. Box 1010
Ruston, LA 71273-1010

Rheem Water Heaters
5780 Peachtree-Dunwoody Rd. N.E.
Atlanta, GA 30342

The Ridge Tool Co.
400 Clark St.
Elyria, OH 44036

Rockwell International
400 North Lexington Ave.
Pittsburgh, PA 15208

Spartan Tool Division
1506 W. Division St.
Mendota, IL 61342

Speakman
P.O. Box 191
Wilmington, DE 19899-0191

Structural Fibers
920 Davis Rd.
Elgin, IL 60123

Symmons Industries, Inc.
31 Brooks Dr.
Braintree, MA 02184

Thompson Plastics, Inc.
3425 Stanwood Blvd. N.E.
Huntsville, AL 35811

Universal-Rundle Corp.
217 N. Mill St.
New Castle, PA 16103

Vanguard Plastics
831 N. Vanguard St.
McPherson, KS 67460

Water Conditioner, Inc.
509 W. Main St.
P.O. Box 187
Waunakee, WI 53597

Woodford Manufacturing Co.
2121 Waynoka Rd.
Colorado Springs, CO 80915

Zoeller Co.
3280 Old Millers Lane
Louisville, KY 40216

Index